P9-DWH-298

HILLSBORO PUBLIC LIBRARIES
Hillsboro, OR
Member of Washington County
COOPERATIVE LIBRARY SERVICES

THE HALF-LIFE OF FACTS

CURRENT

THE
HALF-LIFE
OF
FACTS

Why Everything
We Know
Has an
Expiration Date

SAMUEL ARBESMAN

HILLSBORO PUBLIC LIBRARIES
Hillsboro, OR
CURRENT
Member of Washington County
COOPERATIVE LIBRARY SERVICES

CURRENT
Published by the Penguin Group
Penguin Group (USA) Inc., 375 Hudson Street,
New York, New York 10014, U.S.A.
Penguin Group (Canada), 90 Eglinton Avenue East, Suite 700,
Toronto, Ontario, Canada M4P 2Y3
(a division of Pearson Penguin Canada Inc.)
Penguin Books Ltd, 80 Strand, London WC2R 0RL, England
Penguin Ireland, 25 St. Stephen's Green, Dublin 2, Ireland
(a division of Penguin Books Ltd)
Penguin Books Australia Ltd, 250 Camberwell Road, Camberwell,
Victoria 3124, Australia
(a division of Pearson Australia Group Pty Ltd)
Penguin Books India Pvt Ltd, 11 Community Centre, Panchsheel Park,
New Delhi – 110 017, India
Penguin Group (NZ), 67 Apollo Drive, Rosedale, Auckland 0632,
New Zealand (a division of Pearson New Zealand Ltd)
Penguin Books (South Africa) (Pty) Ltd, 24 Sturdee Avenue,
Rosebank, Johannesburg 2196, South Africa

Penguin Books Ltd, Registered Offices:
80 Strand, London WC2R 0RL, England

First published in 2012 by Current,
a member of Penguin Group (USA) Inc.

10 9 8 7 6 5 4 3 2 1

Copyright © Samuel Arbesman, 2012
All rights reserved

LIBRARY OF CONGRESS CATALOGING IN PUBLICATION DATA
Arbesman, Samuel.
  The half-life of facts : why everything we know has an expiration date /
Samuel Arbesman.
    pages ; cm
  Includes bibliographical references and index.
  ISBN 978-1-59184-472-3
  1. Evolution.   2. Science—Philosophy.   3. Probabilities.   I. Title.
  Q175.32.E85A74 2012
  501—dc23
                                                  2012019142

Printed in the United States of America
Set in Sabon
Designed by Spring Hoteling

No part of this book may be reproduced, scanned, or distributed in any printed or elec-
tronic form without permission. Please do not participate in or encourage piracy of copy-
righted materials in violation of the author's rights. Purchase only authorized editions.

While the author has made every effort to provide accurate telephone numbers, Internet
addresses, and other contact information at the time of publication, neither the publisher
nor the author assumes any responsibility for errors, or for changes that occur after publication. Further, the publisher does not have any control over and does not assume any
responsibility for author or third-party Web sites or their content.

ALWAYS LEARNING   PEARSON

TO DEBRA

CONTENTS

THE HALF-LIFE OF FACTS

CHAPTER 1
The Half-life of Facts

WHEN my grandfather was in dental school in the late 1930s, he was taught state-of-the-art medical knowledge. He learned all about anatomy, many aspects of biochemistry, and cell biology. He was also taught the number of chromosomes in a human cell. The problem was, he learned that it was forty-eight. Biologists had first visualized the nuclei of human cells in 1912 and counted these forty-eight chromosomes, and it was duly entered into the textbooks. In 1953, a well-known cytologist—someone who studies the interior of cells—even said that "the diploid chromosome number of 48 in man can now be considered as an established fact."

But in 1956, Joe Hin Tjio and Albert Levan, two researchers working at Memorial Sloan-Kettering Cancer Center in New York and the Cancer Chromosome Laboratory in Sweden, decided to try a recently created technique for looking at cells. After counting over and over, they nearly always got only forty-six chromosomes. Previous researchers, who Tjio and Levan spoke with after receiving their results, turned out to have been having similar problems. These other scientists had even stopped some of their work prematurely, because they could only find forty-six out of the forty-eight chromosomes that they knew had to be there. But Tjio and Levan didn't make the same assumption. Instead, they made the bold suggestion that everyone else had been using the wrong number: There are only forty-six chromosomes in a human cell.

Facts change all the time. Smoking has gone from doctor recommended to deadly. Meat used to be good for you, then bad to eat, then good again; now it's a matter of opinion. The age at which women are told to get mammograms has increased. We used to think that the Earth was the center of the universe, and our planet has since been demoted. I have no idea any longer whether red wine is good for me. And to take another familial example, my father, a dermatologist, told me about a multiple-choice exam he took in medical school that included the same question two years in a row. The answer choices remained exactly the same, but one year the answer was one choice and the next year it was a different one.

Other types of facts, ones about our surroundings, also change. The average Internet connection is far faster now than it was ten years ago. The language of science has gone from Latin to German to English and is certain to change again. Humanity has progressed from a population of less than two billion to more than seven billion people in the past hundred years alone. We have gone from being earthbound to having had humans walk on the moon, and we have sent our artifacts beyond the boundaries of our solar system. Chess, checkers, and even *Jeopardy!* have gone from being the domains of human experts to ones of computerized mastery.

Our world seems to be in constant flux. With our knowledge changing all the time, even the most informed people can barely keep up. All this change may seem random and overwhelming (Dinosaurs have feathers? When did that happen?), but it turns out there is actually order within the shifting noise. This order is regular and systematic and is one that can be described by science and mathematics.

Knowledge is like radioactivity. If you look at a single atom of uranium, whether it's going to decay—breaking down and unleashing its energy—is highly unpredictable. It might decay in the next second, or you might have to sit and stare at it for thousands, or perhaps even millions, of years before it breaks apart.

But when you take a chunk of uranium, itself made up of trillions upon trillions of atoms, suddenly the unpredictable becomes predictable. We know how uranium atoms work in the aggregate. As a group of atoms, uranium is highly regular. When we combine particles together, a rule of probability known as the law of large numbers takes over, and even the behavior of a tiny piece of uranium becomes understandable. If we are patient enough, half of a chunk of uranium will break down in 704 million years, like clockwork. This number—704 million years—is a measurable amount of time, and it is known as the half-life of uranium.

It turns out that facts, when viewed as a large body of knowledge, are just as predictable. Facts, in the aggregate, have half-lives: We can measure the amount of time for half of a subject's knowledge to be overturned. There is science that explores the rates at which new facts are created, new technologies are developed, and even how facts spread. How knowledge changes can be understood scientifically.

This is a powerful idea. We don't have to be at sea in a world of changing knowledge. Instead, we can understand how facts grow and change in the aggregate, just like radioactive materials. This book is a guide to the startling notion that our knowledge— even what each of us has in our head—changes in understandable and systematic ways.

HOW exactly is knowing how knowledge changes actually useful? You may find it interesting to discover that the dinosaurs of our youth—slow, reptilian, and gray-green—are now fast moving, covered in feathers, and the colors of the NBC peacock. But if you don't have a six-year-old at home, this is probably not going to affect your life in any significant way.

I could tell you that certain areas of medical knowledge have a churn of less than a half century—well within a single life span—and knowing this can motivate us to constantly brush up on what we know, so we continue to eat healthy or exercise correctly and don't simply rely on what we were told when we were young. Or I could say that by knowing how language changes, we can better understand the slang and dialect of the generation that follows us.

But really, practical examples like these, while important (medical knowledge will keep cropping up), are not the main point. Knowing how facts change, how knowledge spreads, or how we adapt to new ideas are all important for a different reason: Knowing how knowledge works can help us make sense of our world. And even more than that, it can allow us to anticipate the shortcomings in what we each might know and help us to plan for these flaws in our knowledge.

Facts are how we organize and interpret our surroundings. No one learns something new and then holds it entirely independent of what they already know. We incorporate it into the little edifice of personal knowledge that we have been creating in our minds our entire lives. In fact, we even have a phrase for the state of affairs that occurs when we fail to do this: cognitive dissonance.

Ordering our surroundings is the rule of how we as humans operate. In childhood we give names to our toys, and in adulthood we give names to our species, chemical elements, asteroids, and cities. By naming, or, more broadly, by categorizing, we are creating an order to an otherwise chaotic and frightening world.

And when we learn facts, we are doing the same thing. Facts—whether about our surroundings, the current state of knowledge, or even ourselves—provide us with a sense of control and a sense of comfort. When we see something out of the corners of our eyes around dusk, we needn't view it as a creepy bird of the night: We call that a *bat*, which is a winged nocturnal mammal that "sees" through the use of echolocation and is probably afraid of the bipedal mammals around it. Only half as scary, right?

But when facts change, we lose a little bit of this control. Suddenly things aren't quite as they seemed. If doctors didn't know that smoking was bad for us for decades, we worry about what else doctors are also wrong about today. If I've just learned that my parents had completely different—and, for their time, acceptable—parenting techniques from my own, I am a bit concerned about my upbringing. And if I've just found out that scientists have discovered hundreds of planets outside the solar system, and I was living with the assumption that there were only a handful, I might be a little shaken, or at the very least somewhat surprised.

But if we can understand the underlying order and patterns of how facts change, we can better handle all of the uncertainty that's around us.

To be clear: I'm using the word *fact* in an intuitive way—a bit of knowledge that we know, either as individuals or as a society, as something about the state of the world. We generally like our facts to represent an accurate representation of reality, an objective truth, but that's not always the case.

Certain fields use fact to mean an objective truth. The endeavor of science gets us ever closer to this truth, and many of the shifts in what we know occur only at the fringes of discovery, and are due to our continuous approach to truth. However, I am choosing to use *fact* in a looser fashion, simply to refer to our individual states of knowledge awareness. This can refer to a scientific fact, even if it might eventually be disproved, or even to a less ambiguous sort of fact, such as the current fastest human runner or the most powerful computer, which are facts about our surroundings.

This will not satisfy everyone, but there are two reasons for this choice. The first is to avoid being drawn down some sort of epistemological rabbit hole, which is likely to be more than a little confusing. To paraphrase Justice Potter Stewart, many of us know a fact when we see one. But second, and more important, it turns out that lots of different types of knowledge change in similar ways. While some facts are about approaching truth and some are about our changing surroundings, we can only see the similarities clearly in how they operate by bundling all of these types of facts together.

And there's one simple way to organize facts, even before we understand all the math and science behind how knowledge changes. We can organize what we know by the rate at which it changes.

Imagine we have all the facts in the world—those pieces of knowledge that contain all that we know—lined up according to how often they change. On the far left we have the fast-changing facts, the ones that are constantly in flux. These are things such as what the weather will be tomorrow or what the stock market close was yesterday. And on the far right we have the very slow-changing facts, the ones that for all practical purposes are constant. These are facts such as the number of continents on the planet or the number of fingers on a human hand.

In between we have the facts that change, but not too quickly—and are therefore that much more maddening to deal with. These facts might change over the course of years or decades or a single lifetime. How many billions of people are on the planet is one of these. I learned five billion in school, and we just recently crossed seven billion, as of 2012. My grandfather, who was born in 1917, learned there were fewer than two billion. The number of planets outside the solar system that have been discovered, or, for that matter, the number of planets in our own solar system, is also in this category. What we know about dinosaurs is in this group of facts, as is the average speed of a computer. The vast majority of what we know seems to fall into this category, which I call *mesofacts*—facts that change at the meso-, or middle, timescale.

Lots of our scientific knowledge consist of mesofacts. For example, the number of known chemical elements is a mesofact. If, as a baby boomer, you learned high school chemistry in 1970, and then, as we all are apt to do, did not take care to brush up on your chemistry periodically, you would not realize that there are at least 12 new elements in the periodic table, bringing the total up to 118. Over a tenth of the elements have been discovered since you graduated high school. And all that dinosaur knowledge I've mentioned is also made up of mesofacts.

Technology is full of mesofacts too, from the increase in

transportation speeds to the changes in how we store information—from floppy disk to the cloud. The height of the tallest skyscraper has also steadily increased over time due to improving technology.

World records are constantly being broken in the realms of human ability, and we recently have begun to be humbled by machines, as they rack up wins against us in more and more games once deemed too complex for computers, from Othello and checkers to chess. All mesofacts.

Mesofacts are all around us and just acknowledging their existence is useful. It can help eliminate part of the surprise in our lives.

If my grandfather had been told in dental school that a specific fraction of the knowledge he learned there would become obsolete soon after he graduated, this could at least provide an anchor for his uncertainty. It would prevent dentists of his generation from being surprised by basic biological facts, or from working with outdated knowledge. In fact, many medical schools now do this: They embrace the mesofacts of medicine, teaching physicians that changing knowledge is the rule rather than the exception.

But simply knowing that knowledge changes like this isn't enough. We would end up going a little crazy as we frantically tried to keep up with the ever-changing facts around us, forever living on some sort of informational treadmill. But it doesn't have to be this way, because there are patterns: Facts change in regular and mathematically understandable ways. And only by knowing the pattern of our knowledge's evolution can we be better prepared for its change.

THERE are mathematical regularities behind the headlines of changing scientific knowledge. We accumulate scientific knowledge like clockwork, with the result that facts are overturned at regular intervals in our quest to better understand the world. Similarly, the growth and change of technological knowledge, from processing power to information storage, are also part of the universe of facts that change with regularity. And of course these two areas—science and technology—affect many other factual aspects of our lives: from the spread of disease, to how we travel, and even to the increase in computer viruses on the Internet. All of these areas of knowledge change systematically.

But just as the creation of facts operates according to certain scientific principles, so too does the spread of knowledge; how each of us hears of new information, or how error gets dispelled, adheres to the rules of mathematics. And due to a new understanding of cognitive biases, much of what each of us knows, even as it changes, now has a clear structure: These shifts obey certain scientific patterns that are explicable by findings in cognitive science.

This is not to say that we can understand everything. Much in the world is shocking and unexpected, and we still have to deal with these new facts when we become aware of them. But by and large there are ways to understand how our knowledge changes, ways to bring order to the chaos of ever-changing facts.

William Macneile Dixon, a British professor of literature in the late nineteenth and early twentieth centuries, once wrote, "The facts of the present won't sit still for a portrait. They are constantly vibrating, full of clutter and confusion."

We now understand how vibrations work, due to physics. We're no longer confused by the fact that plucking a guitar string somehow gives rise to order and music. It's time we do the same thing for the fluctuations in what we know as well, and recognize that there's an order to all of our changing knowledge. This book is a guide to the science behind the vibrations in the facts around us.

CHAPTER 2
The Pace of Discovery

WHEN Derek J. de Solla Price arrived at Raffles College (now the National University of Singapore) to lecture on applied mathematics in 1947, he did not intend to spearhead an entirely new way of looking at science. But his plans to continue research in physics and mathematics were altered by the construction on the college library. Since Raffles was a small university, the library was actually giving books out to students and faculty to store in their dormitories and apartments while the construction was under way.

Price ended up with a complete set of *Philosophical Transactions* of the Royal Society of London, a British scientific journal that dates back to 1665. Once home, he stacked the journals chronologically against the walls in his apartment: Each pile was published later than the one before it, and they were all lined up one after another. One day, while idly looking at this large collection of books that the library had foisted upon him, he realized that the heights of the piles of these bound volumes weren't all the same. But their heights weren't random either. Instead, he realized, the heights of the volumes fit a specific mathematical shape: an exponential curve. Price's simple observation was the origin of a sophisticated quantitative theory for how scientific knowledge advances.

. . .

MOST of our everyday lives revolve around linear growth, or changes that can be fit onto a line. When something increases by the same amount each year, when the rate is constant, we get linear growth. When we drive somewhere, and go at the same speed the entire way, a chart showing the distance we've traveled over time follows a straight line. And if we have a machine that builds widgets at a constant rate of three per hour, the number of widgets after a given number of hours grows linearly with the number of hours we're considering.

Due to how easy it is to imagine (and our brains seem particularly well suited to this type of thinking), we often think in terms of linear growth. If the temperature was sixty-five degrees yesterday, and sixty degrees the day before that, it is not surprising if we expect it to be about seventy degrees today.

But there are many examples of change that occur differently. If you were watching when the sun set over the course of a few days early in the summer, it wouldn't be unreasonable to expect the sunset's timing to follow a nice linear curve: Each day the sun sets the same number of minutes later than it did the day before. But it turns out that sunsets at a specific location adhere to a sine curve—a wavelike shape that looks like a rope being shaken up and down, a shape that we aren't particularly intuitive about. During the solstices—the shortest and longest days of the year—we are at the top or bottom of the wave, when the sunset only varies by a small amount each day; during the equinoxes (spring and fall), we are in the steep parts of the wave, and each day the sunset time is many minutes different from the day before. This is far from a curve that we can think about easily.

We are just as ill suited when it comes to noticing the many changes that adhere to exponential growth. When we encounter exponential curves all around us, we often don't think about it this way at all, because it is harder to picture. Exponential growth is when something increases by the same fraction or percentage, rather than the same amount, each second or minute or hour. If bacteria double every hour, that's exponential growth, because they're growing at a constant rate of 200 percent an hour. Compound interest is the same sort of thing: If our money grows by a certain percentage each year, we can describe this growth by an exponential curve.

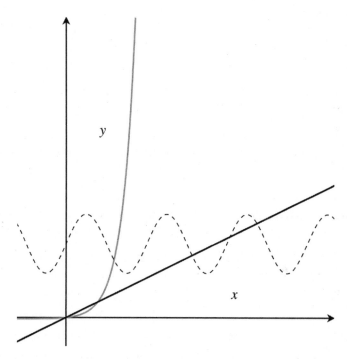

Figure 1. A linear (black) versus exponential (gray) curve versus sine (dotted) curve.

As you might have realized, exponential growth is very rapid. Even if we are initially adding only a small amount to some quantity each hour or day, that quantity can become very big very quickly. Imagine we are given a penny and begin doubling it each day. After a week we would be receiving less than a $1.50 a day. But give it one more week. Now we're getting more than $80 a day. Within a month our allowance is more than $100 million a day!

Exponential growth gets its name from the use of an exponent: an exponent signifies how many times to multiply another number, the base, by itself. Many times a special constant is used for the base; in the case of exponential growth it is often *e*. Also known as Napier's constant, it is about 2.72. It's one of those numbers, like π, that crops up in the weirdest situations, from bacteria doubling to infinitely long sums of numbers. The exponent part of the equation includes what is known as its rate of growth. The larger this value, the faster the quantity grows, and the faster it doubles.

THE exponential growth curve was well-known to Price, so when he began to measure the heights of his stacks of journals, he knew immediately what was going on. But maybe he just happened to have gotten the only stack of journals that obeyed this curious pattern. So he began collecting lots of data, a research style that he followed throughout his life.

He measured the number of journal articles in the physics literature in general, as well as for more specialized fields, such as the subfield that deals with linear algebra. And they all seemed to have elements of the exponential curve. Price began to recognize that this could be a new way to think about how science grows and develops. Price published his findings, under the title "Quantitative Measures of the Development of Science," in a small French journal in 1951, after presenting this work at a conference the previous year in Amsterdam.

No one was interested.

But Price wasn't deterred. He returned to Cambridge and continued to pursue his research in this new field, the quantitative study of science, or *scientometrics*, as it soon became known. This science of science was still quite young, but Price set himself to collecting vast quantities of data to help him understand how science changes.

By the 1960s, he was the foremost authority in this field. He gathered data from all aspects of science and marshaled evidence that enabled him to look at scientific growth as something far from haphazard; this knowledge was subject to regular laws.

Expanding on his initial research on scientific journals, he gathered data for a wide variety of areas that displayed this growth, from chemistry to astronomy. Price calculated the doubling times— how long it takes for something to double, a proportional increase that implies exponential growth—for these components of science and technology, which then can be used as a rough metric for seeing how different types of facts change over time. Here is a selection of these doubling times from his 1963 book *Little Science, Big Science*:

Domain	Doubling Time (in years)
Number of entries in a dictionary of national biography	100
Number of universities	50
Number of important discoveries; number of chemical elements known; accuracy of instruments	20
Number of scientific journals; number of chemical compounds known; memberships of scientific institutes	15
Number of asteroids known; number of engineers in the United States	10

The growth of facts was finally beginning to be subjected to the rigors of mathematics.

PARALLEL to Price's work in the hard sciences, a similar line of research was proceeding in the social sciences. In 1947, a psychologist named Harvey Lehman published a curious little paper in the journal *Social Forces*. Combing through a wide variety of dictionaries, encyclopedias, and chronologies, Lehman set out to count the number of major contributions made in a wide variety of areas of study over the years. He looked at everything from genetics and math to the arts, whether new scientific findings, new theorems, or even new operas produced. What he found in all of these were exponential increases in output over time. But this wasn't only over the previous few decades. Lehman looked at each of these areas over hundreds of years. He examined philosophy over the six hundred years from 1275 to 1875, botany over the three hundred years from 1600 to 1900, and geology over the four hundred years from 1500 to 1900.

Each area was found to have a characteristic rate of increase. Here are doubling times (the number of years it takes for the yearly contributions in these fields to double) from Lehman's findings, along with a few more recent areas examined:

Field	Doubling Time (in years)
Medicine and hygiene	87
Philosophy	77
Mathematics	63
Geology	46
Entomology	39
Chemistry	35
Genetics	32
Grand opera	20

Independently, a number of thinkers were coming to the realization that the growth of knowledge was subject to patterns, and was far from random. Similarly, different types of growth fit different types of knowledge creation. For example, opera is a far faster-changing domain than the sciences. Even though science and opera composition are inherently creative, science is limited by what we can determine about nature. Science can develop only as quickly as we can figure out things about the world. Grand opera, however, is not limited by what is true, only by what is beautiful, and should therefore be able to grow more rapidly, since it doesn't have to be rigorously subjected to experimentation.

In addition, we can see a hint of how more fundamental discoveries grow by comparing them to ones that are more dependent on other areas, which build on work done in other fields. For example, genetics and chemistry, two areas of the basic sciences, proceed at similar rates. On the other hand, medicine and hygiene are much slower, and are also areas that rely on these more basic fields for new discoveries. Perhaps this is a hint that more derivative fields move more slowly compared to the more basic areas of knowledge on which they depend.

Price's and Lehman's efforts showed that looking at how knowledge grows in a systematic way was finally possible, and they unleashed a wave of discoveries.

PRICE'S approach, looking at how science progresses by examining scientific articles and their properties, has proven to be the most successful and fastest-growing area of scientometrics. While scientific progress isn't necessarily correlated with a single publication—some papers might have multiple discoveries, and others might simply be confirming something we already know—it is often a good unit of study.

Focusing on the scientific paper gives us many pieces of data to measure and study. We can look at the title and text and, using sophisticated algorithms from computational linguistics or text mining, determine the subject area. We can look at the authors themselves and create a web illustrating the interactions between scientists who write papers together. We can examine the affiliations of each of the authors and try to see which collaborations between individuals at different institutions are more effective. And we can comb through the papers' citations, in order to get a sense of the research a paper is building upon.

Examining science at the level of the publication can give us all manner of exciting results. A group of researchers at Harvard Medical School looked at tens of thousands of articles published by its scientists and mapped out the buildings on campus where they worked. Through this, they were able to look at the effect that distance has on collaboration. They found exactly what they had assumed but no one had actually measured: The closer two people are, the higher the impact of the research that results from that collaboration. They found that just being in the same building as your collaborators makes your work better.

We can also understand the impact of papers and the results within them by measuring how many other publications cite them. The more important a work is, the more likely it is to be referenced in many other papers, implying that it has had a certain foundational impact on the work that comes after it. While this is certainly an imperfect measure—you can cite a paper even if you disagree with it—much of the field of scientometrics is devoted to understanding the relationship between citations, scientific impact, and the importance of different scientists.

Using this sort of approach, scientometrics can even determine what types of teams yield research that has the highest impact. For example, a group of researchers at Northwestern University found that high-impact results are more likely to come from collaborative teams rather than from a single scientist. In other words, the days of the lone hero scientist, along the lines of an Einstein, are vanishing, and you can measure it.

Citations can also be used as building blocks for other metrics. By examining the average number of times articles in a given journal are cited, we can get what is known as the *impact factor*. This is widely used and carefully considered: Scientists want their papers to be published in journals with high impact factors, as it is good both for their research and influences decisions such as funding and tenure. The journals with the highest impact factors have even penetrated the public consciousness—no doubt due to the highly cited individual papers within them—and include the general science publications such as *Nature* and *Science*, as well as high-profile medical journals such as the *New England Journal of Medicine.*

Scientometrics has even given bragging tools to scientists, such as the *h-index*, which measures the impact of a paper on other researchers. It was created by Jorge Hirsch (and named after himself; notice the *h*) and essentially counts the number of articles a scientist has published that have been cited at least that many times. If you have an h-index value of 45, it means that you have forty-five articles that have each been cited at least forty-five times (though you have likely published many more articles that have been cited fewer times). It also has the side benefit of meaning that you are statistically more likely to be a fellow of the National Academy of Sciences, a prestigious U.S. scientific organization.

It shouldn't be surprising that the field of scientometrics has simply exploded in the past half century. While Price and his colleagues labored by hand, tabulating citations manually and depending on teams of graduate students to do much of the thankless grunt work, we now have massive databases and computers that can take a difficult analysis project and do it much more easily. For example, the h-index is now calculated automatically by many scientific databases (including Google Scholar), something inconceivable in previous decades. Due to this capability, we now have scientometric results about nearly every aspect of how science is done. As we spend billions of dollars annually on research, and count on science to do such things as cure cancer and master space travel, we have the tools to begin to see what sorts of research actually work.

Scientometrics can demonstrate the relationship between money and research output. The National Science Foundation has examined how much money a university spends relative to how many articles its scientists publish. Other studies have looked at how age is related to science. For example, over the past decades, the age at which scientists receive grants from the National Institutes of Health has increased, causing a certain amount of concern among younger scientists.

There's even research that examines how being a mensch is related to scientific productivity. For example, in the 1960s, Harriet Zuckerman, a sociologist of science—someone who studies the interactions and people underlying the entire scientific venture—decided to study the scientific output of Nobel laureates to see if any patterns could be seen in how they work that might distinguish them from their less successful peers. One striking finding was the beneficence of Nobel laureates, or as Zuckerman termed it, *noblesse oblige*. In general, when a scientific paper is published, the author who did the most is listed first. There are exceptions to this, and this can vary from field to field, but Zuckerman took it as a useful rule of thumb. What she found was that Nobel laureates are first authors of numerous publications early in their careers, but quickly begin to give their junior colleagues first authorship. And this happens far before they receive the Nobel Prize.

As one generous Nobel laureate in chemistry put it: "It helps a young man to be senior author, first author, and doesn't detract from the credit that I get if my name is farther down the list." On the other hand, those peers of Nobel laureates who were not as successful tried to maintain first authorship for themselves far more often, garnering more glory for themselves. By their forties, Nobel laureates are first authors on only 26 percent of their papers, as compared to their less accomplished contemporaries, who are first authors 56 percent of the time. Nicer people are indeed more creative, more successful, and even more likely to win Nobel prizes.

These regular patterns of scientists seem evident enough, at least when we look at whole populations of researchers. But what of regularities related to knowledge itself and how it's created? To understand this, we need to begin thinking about asteroids.

ARTHUR C. Clarke was my favorite writer when I was a teenager. His visionary approach to the world around us, and his imaginary futures, were constructed in great detail, and they often pointed to the most positive aspects of humanity. While he is most famous for

2001: A Space Odyssey, he wrote dozens of books and many more essays and short stories.

In one of his other well-known books, *Rendezvous with Rama*, which was published in 1973, a large cylindrical starship flies through our solar system, and a team of astronauts is sent to unravel its secrets. That starship, *Rama*, was initially thought to be an asteroid; however, it is detected by a series of automated telescopes known as SPACEGUARD. These had been put into place after a meteor smashed into northern Italy in 2077, leading humanity to create a sort of early warning system for any objects that might cross Earth's path and potentially threaten our well-being.

When, a number of years after the book's publication, a project like this was actually proposed, its creators paid homage to Clarke and named it the Spaceguard Survey. It uses a variety of discovery methods, including automatic detection of objects in space. In addition to being a real-life incarnation of science fiction, it is also the vanguard of a whole new way of doing science: automatically. One portion of this program, known as Spacewatch, uses automated image processing to detect what are termed NEOs, or near earth objects. In 1992, it was responsible for the first automated discovery of a comet, which now goes by the unwieldy name of C/1992 J1.

Automated science is being done in fields from biology to astronomy to theoretical mathematics, all using computers to make new discoveries and identify new facts. We have projects that find new species by looking at the genetic sequences in water scooped out of the oceans, and ones that allow chemists to automatically discover ways to synthesize specific chemicals. There is even a Web site named TheoryMine to which anyone can go and get a novel mathematical theorem created by a sophisticated computer program that automatically generates a mathematical proof and have it named after you or a loved one. This is automated discovery combined with vanity plates.

But that's not the only big thing going on right now: There is a movement known as citizen science. Everyday individuals are becoming involved in actual scientific discovery. How is this possible?

The principle is rather elegant. While computers are good at lots of things, from adding numbers to counting words in a document, they are often very bad at many simple things: We are still way ahead of computers in labeling photographs or even reading fuzzy or blurred text. This computer limitation, in addition to providing a stumbling block to any robots whose route to global domination relies on caption creation, has created an entirely new field of computer science known as human computation: Simple tasks are given to lots of people to perform, often either for a small amount of money or because someone has cleverly hidden the task in a game. One of the most well-known examples of these are the distorted words we often have to read correctly in order to prove our humanity to a Web site. Rather than simply being an inconvenience, they are now being exploited to actually help digitize such works as the *New York Times* archives. By pairing a distorted known word with one that computers are unable to decipher, everyday users who can read these words are helping bring newspapers and books into digital formats.

Scientists are beginning to use this sort of human computation. These researchers are relying on citizen scientists to help them look through large amounts of data, most of which is too difficult for a computer (or even a single person) to comb through. One example is Galaxy Zoo, in which scientists gave participants pictures of galaxies to help classify them. The participants weren't experts; they were interested individuals who participated in a minutes-long tutorial and were simply interested in space, or wanted to help scientific progress.

Several intrepid scientists turned a fiendishly difficult problem—how to predict what shapes proteins will fold into based on their chemical makeup—into a game. They found that the best players of a simple online game known as Foldit are actually better than our best computers.

Our pattern-detection abilities, and other quirks of how our brains work, still give us the lead in many tasks that are required for new knowledge. So, in truth, the facts that are changing are not changing simply without the involvement of the general population. We are part of the scientific process now. Each of us, not just scientists, inventors, or even explorers, are able to be a part of the process of creating knowledge.

We are living during an auspicious time with respect to knowledge: For the first time, not only has massive computational power allowed for much information related to scientific discovery to be digitized—the amount of scientific data available online for analysis is simply staggering—but discoveries are actually occurring automatically through computational discovery, and also in a widely distributed fashion, through citizen science.

These combined forces, in addition to changing how new scientific knowledge is generated, have enabled us to obtain massive amounts of data on the properties of scientific discoveries and how they are found. This has led to what I, along with one of my collaborators, Nicholas Christakis, have taken to calling *eurekometrics*. Eurekometrics is about studying scientific discoveries themselves. More traditional scientometric approaches that use citations are still very important. They can teach us how scientists collaborate, measure the impact of scientific research, and chart how scientific knowledge grows, but they often tell us nothing about the content of the discoveries themselves, or their properties. For example, rather than looking at the properties of the articles appearing in plant biology journals, we can instead look at the properties of the plant species that have been discovered.

One simple example of eurekometrics—and one that I was involved in—is examining how discoveries become more difficult over time.

IF you look back in history you can get the impression that scientific discoveries used to be easy. Galileo rolled objects down slopes; Robert Hooke played with a spring to learn about elasticity; Isaac Newton poked around his own eye with a darning needle to understand color perception. It took creativity and knowledge (and perhaps a lack of squeamishness or regard for one's own well-being) to ask the right questions, but the experiments themselves could be very simple. Today, if you want to make a discovery in physics, it helps to be part of a ten-thousand-member team that runs a multibillion-dollar atom smasher. It takes ever more money, more effort, and more people to find out new things.

Until recently, no one actually tried to measure the increasing difficulty of discovery. It certainly seems to be getting harder, but how much harder? How fast does it change?

I approached this question in a eurekometric frame of mind and looked at three specific areas of science: mammal species, asteroids, and chemical elements. These areas have two primary benefits: In addition to being from different fields of science, many have clear discovery data going back hundreds of years. The first mammals discovered after the creation of the classification system developed by Carl Linnaeus date to the 1760s. The first asteroid discovered, Ceres, was found in 1801 (and was actually large enough to be thought a planet). And the first modern chemical element discovered (I ignored such elements as lead and gold, which have been known since ancient times) was phosphorus, in 1669.

As I began thinking about how to understand how discoveries get harder, I settled on size. I assumed that size is a good proxy for how easy it is to discover something: The smaller a creature or asteroid is, the harder it is to discover; in chemistry, the reverse is true, and the largest elements are the hardest to create and detect, so I used inverse size. Based on this, I plotted the average ease of discovery over time.

What I found, using this simple proxy for difficulty, was a clear pattern of how discoveries occur: Each dataset adhered to a curve with the same basic shape. In every case, the ease of discovery went down, and in every case, it was an exponential decay.

What this means is that the ease of discovery doesn't drop by the same amount every year—it declines by the same fraction each year, a sort of reverse compound interest. For example, the sizes of asteroids discovered annually get 2.5 percent smaller each year. In the first few years, the ease of discovery drops off quickly; after early researchers pick the low-hanging fruit, it continues to "decay" for a long time, becoming slightly harder without ever quite becoming impossible.

There is even a similarity in one view of medicine. As Tyler Cowen, an economist at George Mason University, has noted, if you tally the number of major advances, or definitive moments, in modern medicine (as chronicled by James Le Fanu) in each decade of the middle of the twentieth century, you get an eventual decline: "In the 1940s there are six such moments, seven moments in the 1950s, six moments in the 1960s, a moment in 1970 and 1971 each, and from 1973 [to] 1998, a twenty-five-year period, there are only seven moments in total."

But here's the wonderful thing: The output of discovery keeps marching on in each of the areas I examined. We keep on discovering more asteroids, new mammals, and increasingly exotic chemical elements, even as each successive discovery becomes harder. These all occur at different rates—we find asteroids much faster than new types of mammal, for example—but we aren't at the end of discovery in any of these areas. In fact, I only know of one area where scientific research has exhausted all discoveries: the "field" of the discovery of new major internal organs.

The trajectory of discovery in human anatomy began in prehistoric times with the discoveries of hearts and lungs, the organs that are rather hard to miss, especially after seeing your colleague disemboweled by a mastodon. This initial flowering of discovery was followed by that of more subtle organs, such as the pituitary gland. But in 1880, a Swedish medical student named Ivar Sandström discovered the parathyroid gland, and the final major internal organ in humans was discovered. That was it. The field's age of discovery was over.

But science as a whole proceeds apace. We pour in more effort, more manpower, and greater ingenuity into further discovery and knowledge. A simple example is one of the first quantities to be studied in the field of scientometrics: the number of scientists over time. The first U.S. PhDs were granted by Yale University in 1861. Since that time the number of scientists in the United States and throughout the world has increased rapidly. For example, the membership of scientists in the Federation of American Societies for Experimental Biology increased from fewer than five hundred in the year 1920 to well over fifty thousand by the late 1990s. This hundredfold increase is extremely rapid in a period of less than eighty years and is indicative of the increase in scientific power through sheer numbers of scientists.

In fact, if you uttered the statement "Eighty percent of all the scientists who have ever lived are alive today" nearly anytime in the past three hundred years, you'd be right. This has allowed more research to be done by larger scientific teams. Not only that, but higher-impact research is done by teams with many more scientists. Of course, growth like this is not sustainable—a long exponential increase in the number of scientists means that at some point the number of scientists would need to exceed the number of humans on Earth. While this may almost be the case on Krypton, Superman's home planet, where the whole population seems to consist entirely of scientists, I don't see this happening on Earth anytime soon. But this rapid growth demonstrates that scientific discovery is by no means anywhere near finished.

For example, in pharmaceutical research, drug companies counter the decreasing number of drugs created per dollar spent by pouring more money into drug discovery. As science grows exponentially more difficult in some areas, affordable technology often proceeds along a similar curve: an exponential increase in computer processing power means that problems once considered hard, such as visualizing fractals, proving certain mathematical theorems, or simulating entire populations, can now be done quite easily. Some scientists arrive at new discoveries without a significant investment in resources by becoming more clever and innovative. When Stanley Milgram did his famous "six degrees of separation" experiment, the one that showed that everyone on Earth was much more closely linked than we imagined, he did it by using not much more than postcards and stamps.

When one area of research becomes difficult, the scientists in that field either rise to the challenge by investing greater effort or shift their focus of inquiry. Physicists have moved into biology, raising new questions that no one had thought to ask before. Mathematicians and computer scientists have turned their formulas and algorithms to the social sciences and unleashed basic new discoveries about the way societies operate. Or scientists figure out ways to make the hard questions much easier, whether by importing techniques from other areas or inventing new methods.

However it happens, scientific discovery marches forward. We are in an exceptional time, when the number of scientists is growing rapidly and consists of the majority of scientists who have ever lived. We have massive collaborative projects, from the Manhattan Project to particle accelerators, that have and are unearthing secrets of our cosmos. Yet, while this era of big science has allowed for the shockingly fast accumulation of knowledge, this growth of science is not unexpected. Both the growth of scientific facts itself as well as what allows discovery and innovation to plow ahead can be explained by scientometrics.

However, just as scientometrics can explain the growth of scientific knowledge, it can also explain how facts are overturned.

CHAPTER 3
The Asymptote of Truth

ANYONE who was alive during the Late Cretaceous, between about one hundred million and sixty-five million years ago, would have seen many familiar creatures: the fearsome tyrannosaurs; duck-billed hadrosaurs; numerous birds; and even a selection of small mammals. If you ventured into the sea, you would have seen a wide variety of marine creatures. Among these were animals known as coelacanths (pronounced "SEE-luh-canths")—gray, thick-lipped fish, they were one of the more hideous creatures of this period. A large meteor struck the Earth at the end of the Cretaceous period around what is today Mexico's Yucatán peninsula, numerous volcanoes erupted that deposited ash across the sky, and the planet was plunged into a cataclysmic climate shift that caused the extinction of all of these creatures. In addition to wiping out our beloved dinosaurs, this massive extinction included the coelacanth, the ugly stepsister to nearly everything else that lived in the ocean.

But in 1938, on the eve of World War II, this changed. While the cause of the massive extinction at the boundary between the Cretaceous and Tertiary periods wasn't yet known, the numerous extinct species from that time had already been chronicled. From the *Tyrannosaurus rex* to the coelacanth, what was lost in the past had been well studied. At the time, Marjorie Courtenay-Latimer, a young woman in South Africa, was curator of a small museum in the town of East London, not far from Cape Town. She had befriended a fisherman in the area, and he would periodically show his catch to her, allowing her to add any possible finds to the museum's collection.

One winter day at the end of 1938, when Courtenay-Latimer went to the dock to check out the fisherman's haul, she noticed a strange fin poking out. When she excavated it from the rest of the pile she discovered what she described as "the most beautiful fish I had ever seen, five feet long, and a pale mauve blue with iridescent silver markings." No doubt this was the scientist in her speaking, with the thrill of seeing something possibly new. Anyone else would have seen a terribly ugly, oily, and foul-smelling fish.

That was definitely the assessment of a taxi driver when Courtenay-Latimer attempted to catch a ride back to the museum with her large, stinking find. But her hunch that this fish was important was verified when she scoured her books back at the museum and identified her catch as the long-lost coelacanth. Somehow, Courtenay-Latimer was astonished to discover, this fish had survived in the Indian Ocean unchanged by evolutionary pressures for tens of millions of years. This was confirmed by a professor at a nearby university, who upon seeing sketches of her find sent the telegram MOST IMPORTANT PRESERVE SKELETON AND GILLS = FISH DESCRIBED.

It took another decade and a half, and the offer of a large reward, for a second coelacanth specimen to be discovered. This one was found off the Comoros islands, between Madagascar and the African mainland. But the impossible had been done: A supposedly extinct species had been discovered alive and well.

The coelacanth is an example of what are known as Lazarus taxa: living things that are presumed long extinct until contrary evidence is discovered. Of course, predicting whether a single extinct species will one day be rediscovered living in some corner of the planet is nearly impossible. But if we look at large groups of species we sometimes can determine, in aggregate, how many species might actually not be extinct after all, and how often facts are incorrect and need to be overturned.

In 2010, two biologists at the University of Queensland in Australia tabulated all the mammals that have very likely gone extinct in the past five hundred years. This yielded a list of 187 species. Then they checked to see how many were eventually recategorized as nonextinct. The answer: More than a third of all mammals that allegedly were lost to time in the past five hundred years have since been rediscovered.

This sort of large-scale analysis is not just for understanding the nature of Lazarus taxa. It can be extended more generally to enable us to understand the entire edifice of science and how we overturn long-held scientific beliefs. By looking at how science changes overall, we can see the patterns in how scientific knowledge is revised. And it can lead us to measuring the half-life of facts.

AS scientific knowledge grows rapidly, it leads to a certain overturning of old truths, a churning of knowledge. While this churning is hard to deny—recall my inability to recall the health benefits of red wine despite having seen it in the newspapers many times—it is difficult to measure. But if we could quantify this churn, that could provide a handle for our uncertainty, and even a metric for how often we should revisit a subject.

A few years ago a team of scientists at a hospital in Paris decided to actually measure this. They decided to look at fields that they specialized in: cirrhosis and hepatitis, two areas that focus on liver diseases. They took nearly five hundred articles in these fields from more than fifty years and gave them to a battery of experts to examine.

Each expert was charged with saying whether the paper was factual, out-of-date, or disproved, according to more recent findings. Through doing this they were able to create a simple chart that showed the amount of factual content that had persisted over the previous decades. They found something striking: a clear decay in the number of papers that were still valid.

Furthermore, they got a clear measurement for the half-life

of facts in these fields by looking at where the curve crosses 50 percent on this chart: forty-five years. Essentially, information is like radioactive material: Medical knowledge about cirrhosis or hepatitis takes about forty-five years for half of it to be disproven or become out-of-date. This is about twice the half-life of the actual radioisotope samarium-151.

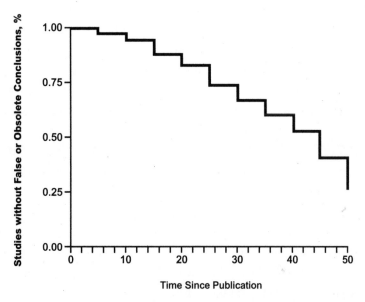

Figure 2. Decay in the truth of knowledge in the areas of hepatitis and cirrhosis. The 50 percent mark is around forty-five years, meaning it takes about forty-five years for half of the knowledge in these fields to be overturned. From Poynard, et al. "Truth Survival in Clinical Research: An Evidence-Based Requiem?" *Annals of Internal Medicine* 136, no. 12 (2002): 888–95.

As mentioned earlier, while each individual radioactive atom's decay is subject to a great deal of uncertainty, in the aggregate, they are far from random. They are subject to a systematic degradation and encapsulated in the shorthand of a single number—the half-life—that denotes how long it takes for half of the material to be subject to radioactive decay.

Knowledge in a field can also decay exponentially, shrinking by a constant fraction. It is like one of Zeno's Paradoxes, according to which we keep getting halfway closer to the finish line but never quite reach it. In this case, the finish line is the point at which no papers from the original batch of cirrhosis and hepatitis studies are still true. While there will always be an infinitesimal number of papers cited many decades, or even centuries, from now, within a certain number of years the vast number of articles will have decayed into irrelevance. Of course, some of these are not wrong, just obsolete. These scientists noted that the effectiveness of treatments in decades past doesn't necessarily become nullified; they simply become superseded by something newer, such as novel vaccines that make treatment of a disease no longer necessary.

But ultimately, while we can't predict which individual papers will be overturned, just like we can't tell when individual radioactive atoms will decay, we can observe the aggregate and see that there are rules for how a field changes over time. In addition, these results are nearly identical to a similar study that examined the overturning of information in surgery. Two Australian surgeons found that half of the facts in that field also become false every forty-five years. As the French scientists noted, all of these results verify the first half of a well-known medical aphorism by John Hughlings Jackson, a British neurologist in the nineteenth and early twentieth centuries: "It takes 50 years to get a wrong idea out of medicine, and 100 years a right one into medicine."

This means that despite the ever-expanding growth of scientific knowledge, the publication of new articles, refutations of existing theories, the bifurcations of new fields into multiple subfields, and the messy processes of grant-writing and -funding in academia, there are measurable ways in which facts are overturned and our knowledge is ever renewed. I'm not simply extrapolating from this half-life of medicine to argue that all of science is like this. Other studies have been performed about the half-life of different types of scientific knowledge as well.

Unfortunately, convening a panel of experts and having them comb through all of science's past conclusions and giving a thumbs-up or thumbs-down to a paper's validity isn't quite feasible. So we have to sacrifice precision for our ability to look at lots and lots of science relatively quickly. One simpler way to do this is by looking at the lifetime of citations. As mentioned before, citations are the coin of the scientific realm and the metric by which we measure the impact of a paper.

Most papers are never cited. And many more are cited only once and then forgotten. Others are only cited by their own authors, in their own other papers. But—and this is no doubt a point in favor of the scientific endeavor—there are numerous papers that are cited by others in the field. And there are the even rarer papers cited so many more times than those around them that they are truly fundamental to a field, towering well above other publications.

To understand the decay in the "truth" of a paper, we can measure how long it takes for the citation of an average paper in a field to end. Whether it is no longer interesting, no longer relevant, or has been contradicted by new research, this paper is no longer a part of the living scientific literature. It is out-of-date. The amount of time it takes for others to stop citing half of the literature in a field is also a half-life of sorts.

This gives us a sense of how knowledge becomes obsolete, but it also has a very practical application. Scholars in the field of information science in the 1970s were concerned with understanding the half-life of knowledge for a specific reason: protecting libraries from being overwhelmed.

In our modern digital information age, this sounds strange. But in the 1970s librarians everywhere were coping with the very real implications of the exponential growth of knowledge: Their libraries were being inundated. They needed ways to figure out which volumes they could safely discard. If they knew the half-life of a book or article's time to obsolescence, it would go a long way to providing a means to avoid overloading a library's capacity. Knowing the half-lives of a library's volumes would give a librarian a handle on how long books should be kept before they are just taking up space on the shelves, without being useful.

So a burst of research was conducted into this area. Information scientists examined citation data, and even usage data in libraries, in order to answer such questions as, If a book isn't taken out for decades, is it that important anymore? And should we keep it on our shelves?

Through this we can begin to see how the half-lives of fields differ. For example, a study of all the papers in the Physical Review journals, a cluster of periodicals that are of great importance to the physics community, found that the half-life in physics is about 10 years. Other researchers have even broken this down by subfield, finding a half-life of 5.1 years in nuclear physics, 6 years for solid-state physics, 5.4 years in plasma physics, and so forth. In medicine, a urology journal has a half-life of 7.1 years, while plastic and reconstructive surgery is a bit more long-lived, with a half life of 9.3 years (note that this is far shorter than the half-life of 45 years calculated earlier, because we are now looking only at citations, not whether something has actually been disproved or rendered obsolete). Price himself examined journals from different fields and found that the literature turnover is far faster in computer science than psychiatry, which are both much faster than fields in the humanities, such as Civil War history.

Different types of publications can also have varied half-lives. In 2008, Rong Tang looked at scholarly books in different fields and found the following half-lives.

Field	Half-life (in years)
Physics	13.07
Economics	9.38
Math	9.17
Psychology	7.15
History	7.13
Religion	8.76

It seems here that physics has the longest half-life of all the fields examined, at least when it comes to books. This is the opposite of what is found in the realm of articles, where the hard sciences are overturned much more rapidly than the social sciences. This could very well be due to the fact that in the hard sciences only the research that has weathered a bit of scrutiny actually makes it into books.

Overall, though, it's clear that some fields are like the radioactive isotopes injected into someone undergoing a PET scan that decay extremely rapidly. Other fields are much more stately, like the radiocarbons, such as carbon-14, used for the scientific dating of ancient artifacts. But overall, these measurements provide a grounding for understanding how scientific facts change around us.

The story of why facts get overturned—sloppy scientists or something else?—is for chapter 8, and has to do with how we do science and how things are measured. But shouldn't the very fact that most scientific knowledge decays be somewhat distressing?

It's one thing to be told that a food is healthy one day and a carcinogen the next. But it's something else entirely to assume that basic tenets of our scientific framework—gravity, genetics, electromagnetism—might very well be wrong and can possibly be part of the half-life of knowledge.

But this is not the way science works. While portions of our current state of science can be overturned, this occurs only in the service of something much more positive: an approach to scientific truth.

IN 1974, three scientists working at the Thermophysical Research Properties Center at Purdue University released a supplement to the *Journal of Physical and Chemical Reference Data*. This was no small undertaking—it was more like an eight-hundred-page book on a single topic: the thermal conductivity of the elements in the periodic table.

Thermal conductivity refers to how easily each element conducts heat. For example, metals are much better conductors of heat than gases (or plastics) are; that's why frying pans often have handles made of plastic instead of metal. But in addition to materials having different inherent thermal conductivities, there are a number of factors that influence these values. One of the most important is temperature. In general, the hotter something already is, the better it is at conducting heat.

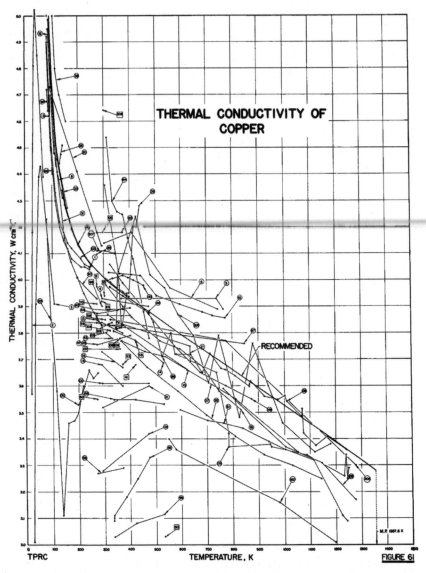

Figure 3. Thermal conductivity of copper versus temperature, as derived from multiple experiments. Reprinted with permission from Ho, et al. "Thermal Conductivity of the Elements." *Journal of Physical and Chemical Reference Data* 1, no. 2 (April 1972): 279–421. © 1972, American Institute of Physics.

This supplement is an exhaustive, data-point-filled text that goes through every chemical element and examines the concept of thermal conductivity. But measuring these curves—trying to determine the relationship between temperature and conductivity for each element—isn't always that easy. Therefore, they compiled lots of previous research that had gone into measuring these properties and, based on that research, tried to determine what this curve truly is.

By conducting lots and lots of measurements, and seeing where the results fall on the curve, we can begin to realize what the true nature of this thermal conductivity curve actually is for each element. It can be seen in the graph: There's a lot of noise and uncertainty. But when enough measurements are taken, a really clear picture of how properties are related emerges (in this case, thermal conductivity and temperature). If a certain fraction of the results were removed, or only the results from one of the hundreds of papers cited in the supplement were looked at, there would be a different, less complete, and inaccurate picture of the relationship between thermal conductivity and temperature.

That's how science proceeds.

It's not that when a new theory is brought forth, or an older fact is contradicted, what was previously known is simply a waste, and we must start from scratch. Rather, the accumulation of knowledge can then lead us to a fuller and more accurate picture of the world around us.

Isaac Asimov, in a wonderful essay, used the Earth's curvature to help explain this:

> [W]hen people thought the earth was flat, they were wrong. When people thought the earth was spherical, they were wrong. But if you think that thinking the earth is spherical is just as wrong as thinking the earth is flat, then your view is wronger than both of them put together.

Clearly, when humans went from believing that the Earth was flat to the belief that the Earth was a sphere, there was a big change in their view. In an entirely unmetaphorical way, the shape of people's thoughts was changed. But, as Asimov explained, in terms of practical usage, the flat Earth perspective is not that wrong. The assumption of a flat Earth includes the concept of no curvature at all, or zero inches of curvature per mile. Due to boats appearing on the ocean from over a flat horizon, it can be seen that the curvature is not actually zero. But, as Asimov calculates, it's not that far off. A sphere the size of the Earth has a curvature of only eight inches per mile. That adds up over the size of the Earth, but it's not that big when you think of it as inches per mile.

An entirely spherical world is not correct either. We in fact exist on a very large object known as an oblate spheroid, which has a curvature that varies between 7.973 and 8.027 inches per mile. Each successive worldview, fact, or theory brings us closer to actually explaining how the world truly works and what the state of our environment is. In the case of the Earth's curvature, each new theory got us closer to the correct amount that the Earth curves below our feet. Or, in a more complex example, this is similar to how Einstein's theory subsumed Newton's results and made them even more general. We can still use Newtonian mechanics for everyday purposes (and, in fact, we almost always do), but Einstein refined our understanding of the world at the edges, such as when we are moving at speeds close to the speed of light.

Sometimes we get things entirely wrong, or not as accurate as we would like. But on the whole, the aggregate collection of scientific knowledge is progressing toward a better understanding of the world around us.

To make this abundantly clear, Sean Carroll, a theoretical physicist at Caltech, wrote a wonderful series on his blog that began with a piece entitled "The Laws Underlying the Physics of Everyday Life Really Are Completely Understood." While he's not saying that everything is known about our everyday existence, including the complex notions of "turbulence, consciousness, the gravitational N-body problem, [and] photosynthesis," what Carroll is arguing is that the fundamental laws that underlie the functioning of subatomic particles at everyday temperatures are well-known:

> A hundred years ago it would have been easy to ask a basic question to which physics couldn't provide a satisfying answer. "What keeps this table from collapsing?" "Why are there different elements?" "What kind of signal travels from the brain to your muscles?" But now we understand all that stuff. (Again, not the detailed way in which everything plays out, but the underlying principles.) Fifty years ago we more or less had it figured out, depending on how picky you want to be about the nuclear forces. But there's no question that the human goal of figuring out the basic rules by which the easily observable world works was one that was achieved once and for all in the twentieth century.

Carroll even lays down, in a single equation, how electrons work in normal, everyday room temperatures. While this is a very general and optimistic example (most of our world is not so easily described by a single equation), this is often how we uncover everything in the environment around us: as part of our pattern of discovery we asymptotically approach the truth. Returning to species, we can see this in action.

In 2010, the Census of Marine Life completed its first decade of work. This project, involving more than two thousand scientists from more than eighty countries, was tasked with chronicling and classifying all living things in the ocean. It involves more than a dozen smaller projects and collaborations with organizations and companies, from NASA to Google.

While they are aware that their work is by no means complete, the team has already produced thousands of scientific papers and discovered well over a thousand new species. A quote from *Science Daily* gives a sense of how unbelievable this is:

> On just two stops in the southeast Atlantic Angola Basin, they found almost 700 different copepod species (99 percent

of them unfamiliar) in just 5.4 square meters (6.5 square yards), nearly twice the number of species described to date in the entire southern hemisphere.

Kevin Kelly refers to this sort of distribution as the "long tail of life." In the media world, a small fraction of movies accounts for the vast amount of success and box office take—these are the blockbusters. The same thing happens on the Internet: a tiny group of Web sites commands most of the world's attention. In the world of urban development, a handful of cities holds a vast portion of the world's population. But these superstars aren't the whole story. While they explain a good fraction of what's out there, there is a long tail of smaller movies or cities that exist and are still important. Understanding how they are distributed can give us a better picture of how the world consumes popular culture or lives in cities.

So too with species and many other discoveries. As mentioned earlier, when a field is young the discoveries come easily, and they are often the ones that explain a lot of what is going on—or, in the case of species, are the really big ones. Many things that we know about are incredibly common or relatively easy to know; they're in the main portion of this distribution. But there are uncountably more discoveries, although far rarer, in the tail of this distribution of discovery. As we delve deeper, whether it's into discovering the diversity of life in the oceans or the shape of the Earth, we begin to truly understand the world around us.

So what we're really dealing with is the long tail of discovery. Our search for what's way out at the end of that tail, while it might not be as important or as Earth-shattering as the blockbuster discoveries, can be just as exciting and surprising. Each new little piece can teach us something about what we thought was possible in the world and help us to asymptotically approach a more complete understanding of our surroundings. Whether it's finding multicellular creatures that can live without oxygen or a shrimp that had been thought dead for sixty million years, both of which were found by participants in the Census of Marine Life project, each new discovery adds to all that we know about the universe, in its rich complexity and diversity.

THERE is an order to how science accumulates and explains everything around us that allows us to construct an intricate and ever-improving theory of our world. But just as the science of science can explain how facts are both created and overturned, it can also lead us to understand how other sorts of facts change. And many of these facts are related to the world of technology.

CHAPTER 4
Moore's Law of Everything

I had my first experience with the Internet in the early 1990s. I activated our 300-baud modem, allowed it to begin its R2-D2–like hissing and whistling, and began to telnet. A window on our Macintosh's screen began filling with text and announced our connection to the computers of the local university through this now antiquated protocol. After exploring a series of text menus, I commenced my first download: a text document containing Plato's *Republic*, via Project Gutenberg. Once I completed this task (no doubt after a significant fraction of an hour), I was ecstatic. I can distinctly remember jumping up and down, celebrating that I had gotten this entire book onto our computer using nothing but the phone lines and a lot of atonal beeping.

It took me almost a decade after this incident to actually get around to reading *The Republic*. By the time I did, the notion that we ever expressed wonder at such a mundane activity as downloading a text document seemed quaint. In 2012, people stream movies onto their computers nightly without praising the modem gods. We have gone from the days of early Web pages, with their garish backgrounds and blinking text, to slick interactive sites using cascading style sheets, JavaScript, and so many other bells and whistles that make the entire experience smooth and multimedia-based. No one thinks any longer about modems or the details of bandwidth speeds. And certainly no one uses the word *baud* anymore.

To understand how much has changed, and how rapidly, during the 1990s, we can look to the *Today* show. At one point in January 1994, Bryant Gumbel was asked to read an e-mail address out loud.

He was at an utter loss, especially when it came to the "a, and then the ring around it." This symbol, @, is second nature for us now, but Gumbel found it baffling. Gumbel and Katie Couric then went into a discussion about what the Internet is. They even asked those off camera, "What is 'Internet' anyway?"

The @ symbol has been on keyboards since the first typewriter in 1885, the Underwood. However, it languished in relative obscurity until people began using it as a separator in e-mail addresses, beginning in 1971. Even then, its usage didn't enter the popular consciousness until decades later. Gumbel's confusion, and our amusement at this situation, is a testament to the rapid change that the Internet has wrought.

But, of course, these changes aren't limited to the Internet. When I think of a 386 processor I think of playing SimCity 2000 on my friend's desktop computer, software and hardware that have both long since been superseded. In digital storage media, I have personally used 5¼-inch floppy disks, 3½-inch diskettes, zip discs, rewritable CDs, flash drives, burnable DVDs, even the Commodore Datasette, and in 2012 I save many of my documents to the storage that's available anytime I have access to the Internet: the cloud. This is over a span of less than thirty years.

Clearly our technological knowledge changes rapidly, and this shouldn't surprise us. But in addition to our rapid adaptation to all of the change around us—which I address in chapter 9—what should surprise us is that there are regularities in these changes in technological knowledge. It's not random and it's not erratic. There is a pattern, and it affects many of the facts that surround us, even ones that don't necessarily seem to deal with technology. The first example of this? Moore's Law.

. . .

WE all at least have heard of Moore's Law. It deals with the rapid doubling of computer processing power. But what exactly is it and how did it come about? Gordon Moore, of the eponymous law, is a retired chemist and physicist as well as the cocreator of the Intel Corporation. He founded Intel in 1968 with Robert Noyce, who helped invent the integrated circuit, the core of every modern computer. But Moore wasn't famous or fabulously wealthy when he developed his law. In fact, he hadn't even founded Intel yet. Three years before, Moore wrote a short paper in the journal *Electronics* entitled, "Cramming More Components Onto Integrated Circuits."

In this paper Moore predicted the number of components that it would be possible to place on a single circuit in the years 1970 and 1975. He argued that growth would continue to increase at the same rate. Essentially, Moore's Law states that the processing power of a single chip or circuit will double every year. He didn't arrive at this conclusion through exhaustive amounts of data gathering and analysis; in fact, he based his law on only four data points.

The incredible thing is that he was right. This law has held roughly true since 1965, even as more and more data have been added to the simple picture he examined. While with more data we now know that the period for doubling is closer to eighteen months than a year, the principle stands. It has weathered the personal computer revolution, the march from 286 to 486 to Pentium, and the many advances since then. Just as in science, we have experienced an exponential rise in technological advances over time: Processing power grows every year at a constant *rate* rather than by a constant amount. And according to the original formulation, the annual rate of growth is about 200 percent.

Moore's Law hasn't simply affected our ability to make more and more calculations more easily. Many other developments occur as an outgrowth of this pattern. When processing power doubles rapidly it allows much more to be possible. For example, the number of pixels that digital cameras can process has increased directly due to the regularity of Moore's Law.

But it gets even more interesting. If you generalize Moore's Law from chips to simply thinking about information technology and processing power in general, Moore's Law becomes the latest in a long line of technical rules of thumb that explain extremely regular change in technology.

What does this mean? Let's first take the example of processing power. Rather than simply focusing on the number of components on an integrated circuit, we can think more broadly. What do these components do? They enable calculations to occur. So if we measure calculations per second, or calculations per second at a given cost (which is the kind of thing that might be useful when looking at affordable personal computers), we can ignore the specific underlying technologies that enable these things to happen and instead focus on what they are designed to do.

Chris Magee set out to do exactly that. Magee is a professor at MIT in the Engineering Systems Division, an interdisciplinary department that defies any sort of simple description. It draws people from lots of different areas—physics, computer science, engineering, even aerospace science. But the common denominator is that all of these people think about complex systems—from traffic to health care—from the perspectives of engineering, management science, and the quantitative social sciences.

Magee, along with a postdoctoral fellow Heebyung Koh, decided to examine the progress we've made in our ability to calculate, or what they termed *information transformation.* They compiled a vast data set of all the different instances of information transformation that have occurred throughout history. Their dataset, which goes back to the nineteenth century, is close to exhaustive: It begins with calculations done by hand in 1892 that clocked in at a little under one calculation a minute. Following that came: an IBM Hollerith Tabulator in 1919 that was only about four times faster; the ENIAC, which is often thought of as the world's first computer, that used vacuum tubes to complete about four thousand calculations per second in 1946; the Apple II, which could perform twenty thousand calculations every second, in 1977; and, of course, many more modern and extremely fast machines.

By lining up one technology after another, one thing becomes clear: Despite the differences among all of these technologies— human brains, punch cards, vacuum tubes, integrated circuits— the overall increase in humanity's ability to perform calculations has progressed quite smoothly and extremely quickly. Put together, there has been a roughly exponential increase in our information transformation abilities over time.

But how does this happen? Isn't it true that when a new technology or innovation is developed it is often far ahead of what is currently in use? And if a new technology's not that much better, shouldn't it simply not be adopted? How can all of these combined technologies yield such a smooth and regular curve? Actually, the truth is far messier but much more exciting.

In fact, when someone develops a new innovation, it is often largely untested. It might be better than what is currently in use, but it is clearly a work in progress. This means that the new technology is initially only a little bit better. As its developers improve and refine it (this is the part that often distinguishes engineering and practical application from basic science), they begin to realize the potential of this new innovation. Its capabilities begin to grow exponentially.

But then a limit is reached. And when that limit is reached there is the opportunity to bring in a new technology, even if it's still tentative, untested, and buggy. This progression of refinement and plateau for each successive innovation is in fact described in the mathematical world as a series of steadily rising *logistic curves*.

This is a variation on the theme of the exponential curve. Imagine bacteria growing in a petri dish. At first, as they gobble the nutrients in the dish, they obey the doubling and rapid growth of the exponential curve. One bacterium divides into two bacteria, two bacteria become four, and eventually, one million becomes two million. But soon enough these bacteria bump up against certain limits. They begin to run out of space, literally bumping up against each other, since the size of the petri dish, though very large in the eyes of each individual bacterium, is far from infinite relative to the entire colony.

Soon the growth slows, and eventually it approaches a certain steady number of bacteria, the number that can be safely held in the petri dish over a long period of time. This amount is known as the *carrying capacity*. The mathematical function that explains how something can quickly begin to grow exponentially, only to slow down until it reaches a carrying capacity, is known as a logistic curve.

Of course, the logistic curve describes lots more than bacteria. It can explain everything from how deer populate a forest to how the fraction of the world population with access to the Internet changes over time. It can also explain how people adopt something new.

When a tech gadget is new, its potential for growth is huge. No one has it yet, so its usage can only grow. As people begin to buy the newest Apple device, for example, each additional user is gained faster and faster, obeying an exponential curve. But of course this growth can't go on forever. Eventually the entire population that might possibly choose to adopt the gadget is reached. The growth slows down as it reaches this carrying capacity, obeying its logistic shape.

These curves are also often referred to as S-curves, due to their stretched S-like shapes. This is the term that's commonly used when discussing innovation adoption. Clayton Christensen, a professor at Harvard Business School, argues that a series of tightly coupled and successive S-curves—each describing the progression and lifetime of a single technology—can be combined sequentially when looking at what each consecutive technology is actually doing (such as transforming information) and together yield a steady and smooth exponential curve, exactly as Magee and Koh found. This is known as linked S-curve theory, and it explains how multiple technologies have been combined to explain the shapes of

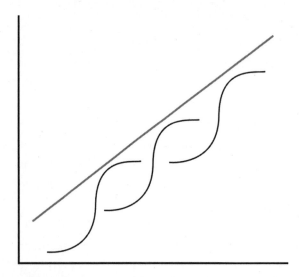

Figure 4. Linked S-curves (or linked logistic curves). When combined, they can yield a smooth curve over time.

change we see over time.

But Magee and Koh didn't simply expand Moore's Law and examine information transformation. They looked at a whole host of technological functions to see how they have changed over the years. From information storage and information transportation to how we deal with energy, in each case they found mathematical regularities.

This ongoing doubling of technological capabilities has even been found in robots. Rodney Brooks is a professor at MIT who has lived through much of the current growth in robotics and is himself a pioneer in the field. He even cofounded the company that created the Roomba. Brooks looked at how robots have improved over the years and found that their movement abilities—how far and how fast a robot can move—have gone through about thirteen doublings in twenty-six years. That means that we have had a doubling about every two years: right on schedule and similar to Moore's Law.

Kevin Kelly, in his book *What Technology Wants*, has cataloged a wide collection of technological growth rates that fit an exponential curve. The doubling time of each kind of technology, as shown in the following table, acts as a sort of half-life for it and is indicative of exponential growth: It's the amount of time before what you have is out-of-date and you're itching to upgrade.

Technology	Doubling Time (in months)
Wireless, bits per second	10
Digital cameras, pixels per dollar	12
Pixels, per array	19
Hard-drive storage, gigabytes per dollar	20
DNA sequencing, dollars per base paid	22
Bandwidth, kilobits per second per dollar	30

Notably, this table bears a strikingly similarity to the chart seen in chapter 2, from Price's research. Technological knowledge exhibits rapid growth just like scientific knowledge.

But the relationship between the progression of technological facts and that of science is even more tightly intertwined. One of the simplest ways to begin seeing this is by looking at scientific prefixes.

IN chapter 8, I explore how advances in measurement enable the creation of new facts and new knowledge. But one fundamental way that measurement is affected is through the tools that we have to understand our surroundings. And we can see the effects of technological advances in measurement by looking at it in one small and simple area: the scientific prefix.

The International Bureau of Weights and Measures, which is responsible for defining the length of a meter, and for a long time maintained in a special vault the quintessential and canonical kilogram, is also in charge of providing the officially sanctioned metric prefixes. We are all aware of *centi-* (one tenth), from the world of length, and *giga-* (one billion), from measuring hard disk space. But there are other, more exotic, prefixes. For example, *femto-* is one quadrillionth and *zeta-* is a sextillion (a one followed by twenty-one zeroes). The most recent prefixes are *yotta-* (10^{24}) and *yocto-* (10^{-24}), both approved in 1991.

But while prefixes are entertaining, and can possibly allow you to win the odd bar bet, the creation of new ones is not just for fun. They are created only when there is a need for them. As technology and science increase exponentially rapidly, so too do the prefix sizes. If you plot prefix sizes against the years they were introduced, you get a roughly exponential progression.

We only measure quantities when we can wrap our scientific minds around them, whether it's measuring energy usage, examining tiny atoms, or thinking about astronomical distances. It would make little sense to have prefixes that referred to numbers in the sextillions, or larger, if we had no use for them. However, as we expand what we know, from the number of galaxies in our universe to the sizes of subatomic particles, we expand our need for prefixes. For example, the cost of genome sequencing is dropping rapidly, even, recently, far faster than exponentially. All of the technological developments facilitate the quick advances of science, and with them the need for new metric prefixes.

These technological doublings in the realm of science are actually the rule rather than the exception. For example, there is a Moore's Law of proteomics, the field that deals with large-scale data and analysis related to proteins and their interactions within the cell. Here too there is a yearly doubling in technological capability when it comes to understanding the interactions of proteins.

Even the field of neuroscience is able to move forward at a pace similar to Moore's Law: The technological advances related to recording individual neurons have been growing at an exponential pace. Specifically, the number of neurons that can be recorded simultaneously has been growing exponentially, with a doubling time of about seven and a half years.

Due to the intermingling of science and technology, how do we disentangle scientific knowledge and technological innovation? Well, sometimes, as we'll see, we can't. This is not to say that there aren't differences, though. As Jonathan Cole, a sociologist of science, argues:

Science and technology are closely related, but they are not the same thing. Science involves a body of knowledge that has accumulated over time through the process of scientific inquiry, as it generates new knowledge about the natural world—including knowledge in the physical and biological sciences as well as in the social and behavioral sciences. Technology, in its broadest sense, is the process by which we modify nature to meet our needs and wants. Some people think of technology in terms of gadgets and a variety of artifacts, but it also involves the process by which individuals or companies start with a set of criteria and constraints and work toward a solution of a problem that meets those conditions.

Henry Petroski, a professor of engineering and history at Duke University, puts it even more succinctly: "Science is about understanding the origins, nature, and behavior of the universe and all it contains; engineering is about solving problems by rearranging the stuff of the world to make new things." Science modifies the facts of what we *know* about the world, while technology modifies the facts of what we can *do* in the world.

Sometimes, though, instead of basic scientific insight leading to new technologies, there are instances where engineering can actually precede science. For example, the steam engine was invented over a hundred years before a clear understanding of thermodynamics—the physics of energy—was developed.

But not only isn't it always clear which one occurs first, it is just as often the case that it's difficult to distinguish between scientific and technological knowledge. Iron's magnetic properties demonstrate this well.

Iron is magnetic, as anyone who has spent any amount of time playing with paper clips and magnets knows. And iron is much more magnetic than aluminum, which you can quickly ascertain by holding up foil to a magnet. These differences in magnetism can be measured, and the amount that a material is magnetic (or not) is known as its *magnetic permeability*.

It turns out that the magnetic permeability of iron has changed over time. Specifically, iron has gotten twice as magnetic every five years. This sounds wrong. Shouldn't the magnetic property of iron be unchanging? Iron is a chemical element, so any amount of this material should be the same, and pure as snow. Why should it instead increase over time?

In truth, the iron that people have used throughout history has actually been far from pure. It has had numerous impurities of all sorts; what could be obtained years ago was far from a perfectly pristine elemental substance. In 1928, the engineer Trygve Dewey Yensen set out to determine the magnetic properties of iron over the previous several decades. By scouring records as far back as 1870, Yensen discovered that iron had steadily, and in a rather exponential fashion, increased its magnetic permeability. And this was entirely due to technology.

As our technological methods for making pure iron have improved, so have the magnetic properties of iron. Something that seems to be safely in the category of scientific fact is actually intimately intertwined with our technological abilities. We have seen a steady and regular shift in these scientific facts as we improved these technologies. But just as technological advances change the scientific facts we already have, new technologies also allow for new discoveries, reflecting the tightly coupled nature of scientific and technological knowledge.

Take the periodic table. The number of known chemical elements has steadily increased over time. However, while in the aggregate the number has grown relatively smoothly, if you zoom in to the data closely, a different picture emerges. As Derek de Solla Price found, the periodic table has grown by a series of logistic curves. He argued that each of these was due to a successive technological advance or approach. For example, from the beginnings of the scientific revolution in the late seventeenth century until the late nineteenth century, more than sixty elements were discovered, using various chemical techniques, including electrical shocks, to separate compounds into their constituent parts. In fact, many of these techniques were pioneered by a single man, Sir Humphry Davy, who himself discovered calcium, sodium, and boron, among many other elements.

However, soon the limits of these approaches became evident, and the discoveries slowed. But, following a Moore's Law–like trajectory, a new technology arose. The particle accelerator was created, and its atom-smashing ability enabled further discoveries. As particle accelerators of increasing energies have been developed, we have discovered heavier and larger chemical elements. In a very real way, these advances have allowed for new facts.

Technological growth facilitates changes in facts, sometimes rapidly, in many areas: sequencing new genomes (nearly two hundred distinct species were sequenced as of late 2011); finding new asteroids (often done using sophisticated computer algorithms that can detect objects moving in space); even proving new mathematical theorems through increasing computer power.

There are even new facts that combine technology with human performance. Athletes break records as their tools—for example, swimsuits, sneakers, and training facilities—become more sophisticated due to technological advances. Even the world of board games has been revolutionized. As noted earlier, over the past several decades, game after game has become a domain where computers dominate, changing the facts around us. Checkers was one of the first ones in which computers were able to beat humans consistently—the computer had its first victory in 1990. Chess and Othello were the next ones people lost to computers, both in 1997, and since 2011 even *Jeopardy!* has become the domain of computer mastery.

Computers can now checkmate better than people, and phrase a correct answer in the form of a question, provinces long thought to be exclusively those of the human mind.

Technology has had a large impact on many other realms of knowledge as well. One that jumps immediately to mind is medicine. Just as our medical knowledge undergoes wholesale changes, so do our medical advances in terms of what is possible. For example, John Wilkins, who in the seventeenth century created what he thought would be the first universal language to help organize our facts and ideas, was himself felled by what is now outdated medical knowledge.

Wilkins likely died due to complications surrounding kidney stones. At the time, the medical options for kidney stones too large to be passed were either (a) a terrible surgery (it involved cutting near the scrotum up into the bladder while the patient was conscious) or (b) a painful death. Wilkins opted out of the surgery, which often killed those who chose it anyway, and died. But medical advances since the Scientific Revolution have progressed such that kidney stones now can be broken up by sound waves, dissolved, or treated otherwise—with high survival rates.

Similarly, medical advances have progressed so rapidly that travelers from previous centuries, if not decades, would scarcely recognize what we have available to us. Not only does a vaccine exist for smallpox, but the disease has been entirely eradicated from the planet. Childbirth has gone from life threatening to a routine procedure. Bubonic plague, far from capable of generating a modern wave of the Black Death, is easily treatable with antibiotics. In fact, when I spent a summer in Santa Fe, we were told that bubonic plague exists in that region not because we should be scared, but just to make our doctors aware of this possibility when we went back to our homes, so they could administer the readily available drugs to treat this scourge of the Middle Ages.

Polio has gone from a menace of childhood summers to a distant memory. A few years ago I was fortunate to attend an exhibit on polio at the Smithsonian National Museum of American History. The disease was presented as something from the history books, and was certainly nothing I had ever experienced, and yet I had a great uncle who walked with a limp due to the disease, and my wife's aunt had it as a child. Reading of people's experiences with the disease, the fear, and the iron lungs was astounding. But through medical advances, polio is now generally regarded in the developed world as a curious artifact of the past.

Technology can even affect economic facts. Computer chips, in addition to becoming more powerful, have gone from prohibitively expensive to disposable. Similarly, while aluminum used to be the most valuable metal on Earth, it plummeted in price due to technological advances that allowed it to be extracted cheaply. We now wrap our leftovers in it.

But occasionally, changes in medical or technological advances don't just alter our lifestyles dramatically, such as in the case of the advent of the Internet. Sometimes they have the potential for fundamentally changing the very nature of humanity. We can see the true extremes of the possibilities of change in the facts of technology by focusing on our life spans.

There has been a rapid increase in the average life span of an individual in the developed world over the past hundred years. This has occurred through a combination of lowered infant mortality and better hygiene, among other beneficial medical and public health practices. These advances have added about 0.4 years to Americans' total expected life spans in each year since 1960. But this increase in life span is itself increasing; it is accelerating.

If this acceleration continues, something curious will happen at a certain point. When we begin adding more than one year to the expected life span—a simple shift from less than one to greater than one—we get what is called *actuarial escape velocity*. What this means is that when we are adding more than one year per year, we can effectively live forever. Let me stress this again: A slight change of the underlying state of affairs in our technological and medical abilities—facts about the world around us—can allow people to be essentially immortal. The phrase *actuarial escape velocity* was popularized by Aubrey de Grey, a magnificently bearded scientist obsessed with immortality. Aubrey de Grey has made the realization of this actuarial escape velocity his life's work.

We're at least several decades from this, according to even the most optimistic and starry-eyed of estimates. And it might very well never happen. But this sort of simple back-of-the-envelope calculation can teach us something: Not only can knowledge change rapidly based on technology, but it can happen so rapidly that it can produce other drastically rapid changes in knowledge. In this case, life spans go from short to long to very long to effectively infinite. Discontinuous jumps in knowledge, and how they occur, are discussed in more detail in chapter 7. But the message is clear: Technological change can affect many other facts, sometimes with the potential for profound change around us.

But what about the opposite direction? Rather than being overly optimistic and assuming massive positive changes in the world based on technology, what about a quantified pessimism? Will we ever reach the end of technology? And are there mathematical regularities here, too?

Just as with science, where naysayers have prognosticated the end of scientific progress, others have done the same with innovation more generally. There is the well-known story of the head of the United States Patent and Trademark Office who said there was nothing more to invent, and a similar story about a patent clerk who even resigned because he felt this to be true.

But there is actually no truth to these stories. In the first case, U.S. Patent Office commissioner Henry Ellsworth, in a report to Congress in 1943, wrote the following: "The advancement of the arts, from year to year, taxes our credulity and seems to presage the arrival of that period when human improvement must end." But Ellsworth wrote this to contrast it with the fact of continuous growth. Essentially, he was arguing that the fact that things continue to grow exponentially, despite the constant feeling that we have reached some sort of plateau, is something startling and worth marveling at. In the other case, the statement by the head of the U.S. Patent Office—that new inventions were things of the past—simply never happened.

However, these stories, and how we use them to laugh at our own ignorance, are indicative of a viewpoint in our society: Not only will innovation continue, but anyone who foresees an end to the growth in technological knowledge is bound to be proven wrong. Technological development, and the changes in facts that go along with it, doesn't seem to be ending anytime soon. Of course, these things must end eventually. The physicist Tom Murphy has shown, in a reductio ad absurdum style of argument, that based on certain fundamental ideas about energy constraints, we will exhaust all the energy in our entire galaxy in less than three millennia. So a logistic curve, with its slow saturation to some sort of upper limit, might be more useful in the long term than a simple exponential with never-ending growth.

In the meantime, technology and science are growing incredibly rapidly and systematically. But there are still questions that need to be addressed: Why do these fields continue to grow? And why do they grow in such a regular manner, with mathematical shapes that are so often exponential curves?

THERE are those who, when confronted with regularities such as Moore's Law, feel that these are simply self-fulfilling propositions. Once Moore quantified the doubling rate of the number of components of integrated circuits, and predicted what would happen in the coming decade, it was simply a matter of working hard to make it come to pass. And once the prediction of 1975 came true, the industry had a continued stake in trying to reach the next milestone predicted by Moore's Law, because if any company ever fell behind this curve, it would be out of business. Since it was presumed to be possible, these companies had to make it possible; otherwise, they were out of the game.

This is similar to the well-known Hawthorne effect, when subjects behave differently if they know they are being studied. The effect was named after what happened in a factory called Hawthorne Works outside Chicago in the 1920s and 1930s. Scientists wished to measure the effects of environmental changes, such as lighting, on the productivity of the workers. They discovered that whatever they did to change the workers' behaviors—whether they increased the lighting or altered any other aspect of their environment—resulted in increased productivity. However, as soon as the study was completed, the productivity dropped.

The researchers concluded that the observations themselves were affecting productivity and not the experimental changes. The Hawthorne effect was defined as "an increase in worker productivity produced by the psychological stimulus of being singled out and made to feel important." While it has been expanded to mean any change in response to being observed and studied, the focus here on productivity is important for us: If the members of an industry know that they're being observed and measured, especially in relationship to a predicted metric, perhaps they have an added incentive to increase productivity and meet the metric's expectations.

But this doesn't quite ring true, and in fact it isn't even possible. These doublings have been occurring in many areas of technology well before Moore formulated his law. As noted earlier, this regularity just in the realm of computing power has held true as far back as the late nineteenth and early twentieth centuries, before Gordon Moore was even born. So while Moore gave a name to something that had been happening, the phenomenon he named didn't actually create it.

Why else might everything be adhering to these exponential curves and growing so rapidly? A likely answer is related to the idea of cumulative knowledge. Anything new—an idea, discovery, or technological breakthrough—must be built upon what is known already. This is generally how the world works. Scientific ideas build upon one another to allow for new scientific knowledge and technologies and are the basis for new breakthroughs. When it comes to technological and scientific growth, we can bootstrap what we have learned before toward the creation of new facts. We must gain a certain amount of knowledge in order to learn something new.

Koh and Magee argue that we should imagine that the magnitude of technological growth is proportional to the amount of knowledge that has come before it. The more preexisting methods, ideas, or anything else that is essential for making a certain technology just a little bit better, the more potential for that technology to grow.

What I have just stated can actually be described mathematically.

An equation in which something grows by an amount proportional to its current size gets exactly what we hoped for: exponential growth. What this means is that if technology is essentially bootstrapping itself, much as science does, and its growth is based on how much has come before it, then we can easily get these doublings and exponential growth rates. Numerous researchers have proposed a whole variety of mathematical models to explain this, using the core idea of cumulative knowledge.

So while exponential growth is not a self-fulfilling proposition, there is feedback, which leads to a sort of technological imperative: As there is more technological or scientific knowledge on which to grow, new technologies increase the speed at which they grow.

But why does this continue to happen? Technological or scientific change doesn't happen automatically; people are needed to create new ideas and concepts. Therefore, in addition to knowledge accumulation, we need to understand another piece that's important to the growth of knowledge: population growth.

SOMEWHERE between ten thousand and twelve thousand years ago, a land bridge between Australia and Tasmania was destroyed. Up until that point individuals could easily walk between Australia and what became this small island off the southern coast of the mainland. Soon after the land bridge vanished, something happened: The tiny population of Tasmania became one of the least technologically advanced societies on the planet.

By the time European explorers came to Tasmania in the seventeenth century, the Tasmanians had only twenty-four distinct devices, as classified by anthropologists, in their toolkit. These twenty-four included such basics as rocks and clubs. In contrast, the Aborigines not far across the strait had hundreds more elements of technology: fishing nets, boats, barbed spears, cold-weather clothing, and much more.

The Tasmanians either never invented these technologies or simply lost them over the millennia.

Joseph Henrich, an anthropologist, constructed a mathematical model to account for how such a loss of technology, or even such a long absence of innovation, could have occurred. The model ultimately comes down to simple numbers. Larger groups of interacting people can maintain skills and innovations, and in turn develop new ones. A small group doesn't have the benefit of specialization and idea exchange necessary for any of this to happen.

Imagine a small group of randomly chosen people stranded on a desert island. Not only would they have just a small subset of the knowledge necessary to re-create modern civilization—assuming Gilligan's professor wasn't included—but only a tiny fraction of the required skills could be done by each person. Much like the economic concept of division of labor, even if we each have two or three skills, to perform all of them adeptly, and also pass them along to our descendants, is a difficult proposition. The maintenance and creation of cultural knowledge are much more easily done with large groups of people; each person can specialize and be responsible for a smaller area of knowledge.

In fact, many economists argue that population growth has grown hand in hand with innovation and the development of new facts. The George Mason University economist Bryan Caplan writes:

> The more populous periods of human history—most obviously the last few centuries—clearly produced more scientific, technological, and cultural innovations than earlier, less populous periods. More populous countries today produce many more scientific, technological, and cultural innovations than less populous countries.

A classic paper by economist Michael Kremer argues this position, in an incredibly sweeping and magnificent article: "Population Growth and Technological Change: One Million B.C. to 1990."

Such a timescale is not for the weak-kneed. In an analysis

worthy of someone as well traveled as Doctor Who, Kremer shows that the growth of human population over the history of the world is consistent with how technological change happens.

Kremer does this in an elegant way, making only a small set of assumptions. First he states that population growth is limited by technological progress. This is one of those assumptions that has been around since Thomas Malthus, and it is based on the simple fact that as a population grows we need more technology to sustain the population, whether through more efficient food production, more efficient waste management, or other similar considerations.

Conversely, Kremer also states that technological growth should be proportional to population size. If invention occurs at the same rate for each person, the more people there are, the more innovation there should be. More recent research, however, shows that population density often causes innovation to grow faster than population size, so this seems like an underestimate. But let's see where Kremer's math takes us.

Using these two assumptions, and a bit of related math, Kremer found that a population's growth rate will increase in size proportionally to the current number of people. To be clear: This is much faster than exponential growth, the fastest growth rate we've considered so far. Exponential growth is a constant rate, and here the rate is growing, and growing along the speed at which the population increases. This is known as a hyperbolic growth rate, and if left unchecked can even result in infinite growth.

Kremer found that until very recently, over the long sweep of human history, this result seems to be true and could be the cause of the rapid technological progress around us. The number of humans in the world has grown in proportion to the current level of the population—the larger the number of people on Earth, the faster the rate at which the population rises.

Furthermore, he found that his model fits with other aspects of world history. For example, just as Tasmania was disconnected from Australia about ten thousand years ago, a number of other land bridges were also destroyed, leading to several populated but disconnected regions. The largest by far was the Old World, which consisted of Europe, Asia, and Africa. Next in size were the Americas, followed by Australia, Tasmania, and Flinders Island, a tiny island off the coast of Tasmania.

And as Kremer predicted, the largest areas—meaning those capable of supporting the largest populations—were the most technologically advanced. The Old World, with its gunpowder and other technologies, led the pack. In second place came the Americas, which were dotted with massive cities, used sophisticated calendars, and had well-developed agriculture. On the other hand, Australian Aborigines remained hunter-gatherers, and Tasmania, as mentioned before, was without even some of the most basic of technologies.

Last we have little Flinders Island, where evidence indicates that the population vanished only four thousand years after its land bridge was destroyed, possibly due to a *technological regress*. This is the phrase Kremer used to signify the loss of even the technologies basic for survival.

But is population really the only story? Or is something more complex going on?

In physics, a simple model that explains the largest amount of the system being studied is often termed a *first-order model*. The more "orders" that are added, the more precise the model will be, as this terminology is derived from the history of fitting functions to complex curves on a graph. The first order explains the general shape, the second order explains a bit of its wiggle, and so on. While each successive term—a higher order—makes the overall model more precise, they each individually explain less and less of the shape of the curve. The first-order model explains most of what's going on, while the higher orders explain the details.

Very likely, population is part of the first-order model of technological progress; it certainly seems that technology and population have gone hand in hand for millennia. However, we know that the likelihood of someone being innovative is not independent of population, as Kremer assumed, and we also know that higher population densities in certain regions need not lead to higher amounts of innovation.

Similarly, it's not just the size of the population that's important, but its parts; the makeup of the population can have an effect on how our facts change. Robert Merton, a renowned sociologist of science, argued in "Science, Technology, and Society in Seventeenth-Century England" that the concerns of the English people during this time period affected where the scientists and engineers of that century focused their attentions. It is unsurprising that they were obsessed with the construction of precise timepieces—that is what was needed in order to carefully measure longitude on the high seas, something of an English preoccupation during this time.

In addition, Merton argued that it wasn't just the overall population size that caused innovation, but who these people were: It turns out that a greater percentage of eminent people of that time chose to become scientists rather than officers of the church or to go into the military. This in turn influenced the rapid innovation of England, rather than overall population size.

The world's evolving technologies and changing facts are not just due to churning out babies and waiting for advances that are due to population growth. New knowledge and innovative technologies are due to a whole host of factors, from the concerns of the populace to the makeup of the population. But to ignore population growth as an important factor for technological innovation is to miss a significant piece of the puzzle.

We've examined technological change and how it's mathematically regular and, even more so, often predictable. We now have a handle on why innovation fits the particular shapes that we see around us. And it's clear that technological change can itself lead to widespread change of other facts. But there is one large area of technology that not only obeys reliable trajectories but also plays a significant role in the spread of other facts and pieces of knowledge: travel and communication.

. . .

DAVID Bradley, a British epidemiologist, decided in 1989 to make a special sort of map. He was interested in the nature of contagion and wanted to see how far people could actually spread a pathogen.

He used data from his own family. He plotted the lifetime distances traveled by the men in his family over four generations: his great-grandfather, grandfather, father, and himself. His great-grandfather only traveled around the village of Kettering, which is north of London, in the county of Northamptonshire. His movements can be encompassed in a square that is about 25 miles on each side. His grandfather, however, traveled a good deal farther, even going so far as London. All of his travels over his lifetime can be defined by a square that is about 250 miles on each side. Bradley's father was even more cosmopolitan and traveled throughout the continent of Europe, leading his lifetime movements to be spread throughout a space that is about 2,500 miles on each side. Bradley himself, a world-famous scientist, traveled across the globe. While the Earth is not a square grid, he traveled in a range that is around 25,000 miles on a side, about the circumference of the Earth. A Bradley man could move ten times farther throughout the course of his life with each successive generation, traveling in a space an order of magnitude more extensive in each direction than his father.

This increase in travel is an exponential increase in distance from one generation to the next. If we look at the areas and not just the distance of the geographic footprint of each man, these also increase exponentially, at a rate double that of the increase in distance (because they are squares). Bradley was concerned with the effect that this increase in travel would have on the spread of disease, postulating that increased travel correlates with an increased spread of disease.

But the Bradley family's exponentially increasing travel distances illustrates not only advances in technology; it is indicative of how technology's march can itself allow for the greater dispersal of other knowledge. What is true of the men in David Bradley's family is true of travel more generally: The speed at which individuals, information, and ideas can spread has greatly increased in the past several hundred years. And, unsurprisingly, it has done so according to mathematical rules.

For example, the upper limit of French travel distances in a single day has obeyed an exponential increase over a two-hundred-year period, mirroring Bradley's anecdotal evidence. Beginning in 1800, as humanity moved from horses to railways, the curve holds. Similar trends hold for air and sea transportation. The curves for sea transport begin a bit earlier (around 1750), and air transit of course starts later (no one is really flying until the 1920s), but like movement over land, these other modes of transportation obey clear mathematical regularities.

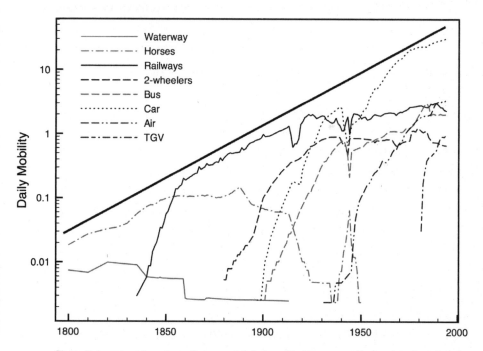

Figure 5. Increases in average distance of daily travel in France over time, using all modes of transportation. Note that the distance traveled is on a logarithmic axis, meaning that the distances capable of being traveled increases exponentially over time. The thick black line shows the general exponential trend. Data from Grübler, *Technology and Global Change* (Cambridge University Press, 2003).

These transportation speeds have clear implications for how the world around us changes.

Cesare Marchetti, an Italian physicist and systems analyst, examined the city of Berlin in great detail and showed that the city has grown in tandem with technological developments. From its early dimensions, when it was hemmed in by the limits of pedestrians and coaches, to later times, when its size ballooned alongside the electric trams and subways, Berlin's general shape was dictated by the development of ever more powerful technologies. Marchetti showed that Berlin's expanse grew according to a simple rule of thumb: the distance reachable by current technologies in thirty minutes or less. As travel speeds increased, so too did the distance traversable and the size of the city. Viewed this way, a city is then a place where people can easily interact.

Furthermore, Bradley's intuition—that transportation speeds are important for understanding the spread of disease—is exactly correct. Just as people can spread at certain rates, so can disease. The Black Death spread precisely at the rate of human movement in the fourteenth century in medieval Europe.

These examples are not exceptions. We arrive at the foundations of a variety of ever-changing facts based on the development of travel technologies: the natural size of a city; how long information takes to wing its way around the world; and how distant a commute a reasonable person might be expected to endure. All of these facts, ever changing, are subject to the rules of technological change. Ultimately, each often follows its own mini–Moore's Law.

FROM communication and urban growth to information processing and medical developments, the facts of our everyday lives are governed by technological progress. While the details of each technological development might be unknown—Will we use disks or CDs? How will we cram more transistors into a square inch?—there are mathematically defined, predictable regularities to how these changes occur. Once we understand this, especially in tandem with understanding scientific progress, we can grasp how technology alters the knowledge around us.

But how exactly do facts spread? And how does this affect how our knowledge changes over time? Just as the technologies of travel and communication affect certain facts of our world, so too do they affect how facts spread and reach each one of us, changing our own personal knowledge.

The facts we, as people, know are due to what we are exposed to, and this requires the spread of knowledge. Is the spread of knowledge just as understandable as how knowledge grows and is overturned? To answer this question we can examine a presidential primary.

CHAPTER 5
The Spread of Facts

WHILE campaigning for the Democratic presidential nomination in 1972, George Wallace was shot multiple times in the abdomen by Arthur Bremer. Wallace, the governor of Alabama, had up until that point been doing very well in the polls. This assassination attempt (he survived, though he was left paralyzed) effectively brought his campaign to an end and altered the election, leaving McGovern to capture the Democratic nomination.

On that same day—May 15, 1972—a group of telephone interviewers happened to be undergoing preparation for that day's assignment at the Consumer Research Corporation, a small market research firm. When David Schwartz, the firm's owner, heard the news of the shooting, he realized this was a rare opportunity: They could use the assassination attempt to actually measure how long it takes for important news to travel and spread through a population. He redirected some of the phone-bank interviewers to examine this, and his team began dialing individuals in the New York City area, attempting to see how the news spread each hour. They carefully called hundreds of people over the course of several hours, and in doing so extracted a clear mathematical curve of how news diffuses over time. Each hour, a larger and larger fraction of those surveyed had heard the news of the shooting. By 10:00 P.M. that night, nearly everyone they spoke with had already heard the news, through a combination of radio, television, and personal contacts. This important piece of information spread extremely rapidly but not instantaneously. The news flashed around New York City in a measurable and predictable way.

Facts do not always diffuse so rapidly. Consider the case of Mary Tai. In February 1994, Tai authored a paper in the journal *Diabetes Care* entitled "A Mathematical Model for the Determination of Total Area Under Glucose Tolerance and Other Metabolic Curves." On the surface, this appears to be little more than a quantitative approach to understanding certain aspects of metabolism, and an article appropriate for such a specialized journal. But look a little closer, specifically at the first few words of the article's title. Need help? Think about determining the area under a curve. And now think about your math classes from high school and college.

What Tai "discovered," even being so bold as to term it Tai's Model, is integral calculus. Tai was not the first person to discover calculus, no doubt to her great disappointment. Rather, it was first developed in the latter half of the seventeenth century by Isaac Newton and Gottfried Leibniz, more than three hundred years before Tai's diabetes-related calculations. Specifically, Tai rediscovered something known as the trapezoidal rule for calculating the area below a curve, which seems to have been known to Newton. And yet Tai's article passed through the editors and has received well over one hundred citations in the scientific literature.

A number of letters written in response to Tai in a later issue of *Diabetes Care* pointed out that this technique is well-known and available in many introductory calculus textbooks. But this example should allow us to recognize something often forgotten: Despite our technological advancement, and even the advances in the speeds of communication chronicled in the last chapter, in many situations knowledge can spread far slower than we might realize.

The creation of facts, as well as their decay, is governed by mathematical rules. But individually, we don't hear of new facts, or their debunking, instantly. Our own personal facts are subject to the information we receive. Understanding how and why information and misinformation spread or don't spread are just as important when it comes to figuring out how we know what we know. Knowledge doesn't always reach all of us simultaneously, whether we're talking about big new theories or simple incorrect facts—it filters through a population in fits and starts. But there are rules for how facts spread, reach individuals, and change what each of us knows.

This occurs most clearly in science itself.

WHILE it may be an extreme case, Tai's mistake is far from the exception in the world of science. When it comes to science, too often knowledge simply doesn't spread as quickly, or as evenly, as we might expect. Disciplines grow rapidly and ramify; it becomes difficult for any one person to know all that has been discovered in a single area.

There has been rapid growth in interdisciplinary research in the past few decades. Molecular biologists work with applied mathematicians, sociologists work with physicists, economists even work with geneticists. If you can think of two fields, you can think of a way to combine a prefix from one and a suffix from the other in order to get a new discipline. People often flippantly acknowledge that an area is about to undergo a shift if the physicists begin to move into it. They have already begun a steady colonization of biology, economics, and sociology, as evidenced by such newfangled terms as *biophysics, econophysics*, and *sociophysics*.

This trend is a welcome one, on the whole, because it often leads to ideas that are well-known in one field finding wonderful applications in another area, where they have not yet been considered. This can lead to an exciting synthesis of ideas, yielding something new and vibrant. But when multiple areas are linked together only superficially, and knowledge is not truly combined, it occasionally leads to a situation in which someone thinks they've discovered something new, yet they're only re-creating something that has been known for a long time in another field. Tai's experience is an extreme example of this, but smaller examples abound.

My own research, which draws from many different disciplines, has not been immune to this problem. In the fall of 2010, when I was a postdoctoral research fellow at Harvard, I was working with Jukka-Pekka Onnela, a fellow postdoc and currently a professor at Harvard's School of Public Health, on a project involving a large anonymized data set of cell phone network calls from a country in Europe. In addition to knowing who called whom, which was important for understanding social ties, we also had information about the callers' locations down to the level of the cell tower. Using our data we could map a community of callers on a country map.

When you do this sort of mapping you get a nice scatter of points on a grid. As part of our work, we wanted to know whether there were clusters of points on this grid, and if so, how many groups of people were there in the points we were examining. I already knew of many sophisticated ways to cluster data points based on their locations, but you often need to know the number of clusters in advance. For example, if you know that there are three groups of data, these algorithms will take your data and place them into three different groups. But what if you didn't know the number?

Onnela and I began searching online for help with this problem, and while we found a lot about clustering, we didn't find the answer to our problem. At one point in this long process one of us might even have recommended thinking about how to create our own method. Then I came up with a solution: "Why don't we just go down the hall and speak to Alan?" Alan is Alan Zaslavsky, a statistician in the Department of Health Care Policy at Harvard, and someone you can count on to be knowledgeable about all things relating to statistics and mathematics (and most other subjects as well). So we walked down the hall and knocked on his door. He wasn't busy, so we went in for a chat, and within five minutes we had the answer: something called the Akaike Information Criterion, created decades ago, was exactly what we needed.

But if we hadn't done this we possibly would have recapitulated some top-notch work done years ago, and likely would have done it far worse than Akaike himself. As our general body of knowledge increases in general, this problem of how to disseminate facts effectively will only become more acute.

Of course, these problems are not even close to new. Even prior to our days of information overload we were plagued with instances of knowledge spreading slowly, or simply failing to spread. The Battle of New Orleans only occurred because news of the end of the War of 1812 traveled so slowly in the early nineteenth century: The war had been over for two weeks when this battle was fought between British and American soldiers. And the Pony Express even promoted its speed by advertising that it delivered news of Lincoln's election in 1860 to the West Coast in only a little more than seven days.

But why does knowledge spread unevenly? Certainly geography is one component: People within reach of the telegraph knew about Lincoln's election much more quickly than those in California, and the closer you were to Washington, D.C., the sooner you knew about the end of the War of 1812. But does that fully explain it, or are there other factors?

One well-studied case of the spread of knowledge, which itself catalyzed how knowledge spread in general, is also one of the most profound innovations of the past thousand years: the printing press.

THE printing press is one of the foundational technologies of our civilization. Unlike the waterwheel, for example, the spread of which is no doubt interesting as well, the printing press has acted as a catalyst for the further dissemination of knowledge. Much like how the production of an enzyme speeds up other chemical reactions, the printing press facilitated the spread and development of new facts. In its first fifty years of existence alone, the price of books in Europe fell by two thirds, a staggering drop in the cost of any item. But did the printing press itself spread rapidly? And were there any patterns in its spread?

The printing press was invented by Johannes Gutenberg in Mainz around 1440. Despite its groundbreaking nature, and as important as it has been since then, it turns out that it did not immediately spread throughout the world. It did not even spread immediately throughout Europe. Rather, it took decades to diffuse, seeping slowly into the cities surrounding Mainz.

Much of the initial spread occurred throughout Germany and northern Italy, and it wasn't until 1476 that the first printing press was found anywhere in England. But it doesn't take decades for information to spread from Germany to England. Even the Black Death, which people tried actively to stop from spreading, made its way across the European continent and the English Channel far faster, in less than half a decade. Clearly geography is not the only part of the story. So why the delay?

Jeremiah Dittmar, an economics professor at American University, has examined this spread carefully. He looked at how certain cities were affected, and why it went from one place to the next. On the following page are a few maps of cities that demonstrate the spread of the printing press over time.

Why wasn't the spread just based on distance? As people got the news of this technology, couldn't they easily begin implementing it themselves?

It turns out that the printing press is far from simple. The technological innovations that Gutenberg developed were much more than the modification of a wine press and the addition of the idea of movable type. Gutenberg combined and extended a whole host of technologies and innovations from an astonishing number of areas, and that is what made his work so powerful. He used metallurgical developments to create metal type that not only had a consistent look (Gutenberg insisted on this), but type that could be easily cast, allowing whole pages to be printed simply at once. He used chemical innovations to create a better ink than had ever been used before in printing. Gutenberg even exploited the concept of the division of labor by employing a large team of workers, many of whom were illiterate, to churn out books at a rate never before

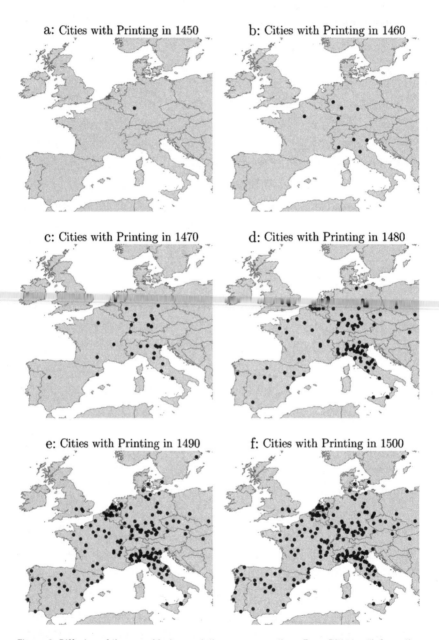

Figure 6. Diffusion of the movable-type printing press over time. From Dittmar. "Information Technology and Economic Change: The Impact of The Printing Press." *The Quarterly Journal of Economics* 126, no. 3 (August 1, 2011): 1133–72, by permission of Oxford University Press.

seen in history. And he even employed elegant error-checking mechanisms to ensure that the type was always set properly: There was a straight line on one side of each piece of type so that the workers could see at a glance whether any letters had been set upside down.

Only by having the combined knowledge of all of these technologies does the printing press become possible and cost-effective. So while it's true that geography—specifically, the distance from Gutenberg's city of Mainz—can explain a great deal of the delay for when certain cities adopted the printing press technology, that's far from the entire story.

The important factor was something more subtle: personal contacts. The cities that got the press first got it because Germans lived there, specifically Germans who had the necessary skills and technologies to make a printing press. These personal contacts allowed for the spread of this semiproprietary technology. If one is from a culture in common with someone else, with common language and traditions, you're more likely to trust each other. This is exactly what happened here. Just as Jews and Huguenots built widespread trading and financial networks, the Germans used their own social ties that had been built on trust and apprenticeship to spread the technologies of the printing press.

This is the rule when it comes to how facts spread: social networks spread information. Of course, back in the day of the printing press, geography and social connectivity were harder to disentangle. As mentioned earlier, just a century prior to Gutenberg, when the Black Death swept across Europe, it spread at the same speed as the rate of movement in that century. But social ties are also vital to the spread of knowledge.

This can be seen by looking at the way the population sizes of cities affected the spread of the printing press. When Dittmar examined city size, he found that larger cities were much more likely to adopt the movable-type printing press technologies soon after their invention, compared to small cities. While only a third of cities in Europe were early adopters, these cities held more than half the population of Europe. Which shouldn't be surprising. Larger cities have more people, yielding more opportunities for there to be a social tie from one city to another. Just as we can look at the German ties, we can see how larger populations mean more ties. Ultimately, when trying to understand the how facts spread, it comes down to social networks.

It's one thing to know that social ties lead to the spread of facts. That is almost intuitively obvious. But are there regularities to how this happens? Can we quantify how facts spread from person to person? Happily, there is an entire field devoted to understanding such networks, which is known as *network science*. Network science examines how connections operate, whether they are connections between people or computers, or even interacting proteins. And just as the mathematics of network science doesn't care what is connected, it is also agnostic about what spreads across these networks. Whether the network is spreading innovations, pieces of news, germs, or pretty much anything else, network science can provide insight.

So it shouldn't be surprising that network science has a great deal to say about the ways in which information and facts can spread, like diseases, from one person to another.

WE are all embedded within social networks. We have friends, neighbors, and relatives. They in turn have contacts of their own. Do this a few more times, and you've reached nearly every person on the planet. That, simply put, is the concept of six degrees of separation.

But knowing the social distance from one individual to another is far from the complete picture. Over the past several decades, network science has developed a far more detailed, though still incomplete, picture of our social interactions. We now understand mathematically why the most popular person in a network has so many more friends than the next most popular person, and we have measured the average number of close social connections each person maintains on a regular basis (it's about four). We understand how social groups are distributed across countries, and even how we make and break friendships over time. In these ways, and many more, we are beginning to truly understand the social structures that we are embedded in, and how these ties influence us.

Some of the most cutting-edge research that is going on right now is devoted to understanding how our connections influence us, and how things spread. As a postdoc I worked in the laboratory of Nicholas Christakis, one of the giants in this field. If you don't recognize his name, you may remember some of the *New York Times* headlines about work done by him and his longtime collaborator James Fowler: "Are Your Friends Making You Fat?"; "Find Yourself Packing It On? Blame Friends"; "Study Finds Big Social Factor in Quitting Smoking"; "Strangers May Cheer You Up, Study Says."

What these researchers have found, in study after study, is that our actions have consequences that ripple across our social web to our friends, our friends' friends, and even our friends' friends' friends.

But just as health behaviors spread, so do facts and bits of knowledge. Since information spreads through social rather than physical space, it is vital that we understand social networks and how they operate. In this globalized age, where we can be anywhere on the planet within a day or so, the ties we have to those we know, rather than where we are, take on greater meaning. Whether we are advertisers trying to gain an advantage in the marketplace, or even just want to lose a few pounds, we crave the answers to a whole host of new kinds of questions about networks. A sample of such questions:

How does each sort of tie that we have to those around us— whether friend, relative, spouse, neighbor—affect the spread of each individual fact or even each behavior? Are our social ties related to distance, which could have an effect on how information spreads? What do the structures of people's social networks look like, and do the shapes of these networks—regular, random, or something in-between—affect how we interact? And are our social ties, such as how many friends we have, and even how likely our friends are to know one another, affected by the genes inside us?

All of these questions are beginning to be asked, and answered, by network scientists. In our specific concern, network scientists have recently begun to explore certain cases where facts spread, or don't spread, and how this works.

BACK in the 1970s a sociologist named Mark Granovetter created a simple little thought experiment: He imagined each social connection between people as having one of two strengths—weak or strong. Strong ties are those that we have to our parents, our spouses, or our close friends. Weak ties are those that we have with friends from high school or college to whom we seldom speak. Or to the acquaintance at work whom we banter with but don't generally speak to outside the office. Or, in the modern age, most of our Facebook "friends."

Granovetter's thought experiment: If we have only these two connection strengths, simplistic though that may be, what should our social networks look like? He argued that if two of our friends are close to us, it is very likely that they will know each other, and probably be close to each other as well. Therefore, much of a social network should consist of clusters of tightly knit groups that are connected by their strong ties into little triangles. But these tight-knit groups are occasionally connected to other strong clusters by weak ties. If these weak ties are the only ties that act as bridges between these little clusters, these weak ties should therefore be very important for facilitating the spread of information far and wide through the network, from one cluster to another. What Granovetter argued for, in other words (and in the words of the title of his celebrated paper), was "The Strength of Weak Ties."

Granovetter even backed this up with some simple data: He

surveyed a group of people on how they got their jobs. Of those who said they got a job through personal contacts, he found that most of these personal contacts were quite "weak."

More recently, scientists have been able to test whether Granovetter was right. Jukka-Pekka Onnela, my former coworker who I mentioned earlier, was actually involved in one of the foundational papers in this area. To understand how information spreads he used a data set that is unbelievably rich and has been the basis for many scientific papers: a collection of anonymized mobile phone calls in a country in Europe.

Using the data about who calls whom, Onnela and his colleagues were able to construct a social network that spans an entire country. But not only did they have the ties between people, they had the strength of each tie: how many minutes people spoke to one another over the course of several months.

They were able to conduct a test using this network: They created an abstract contagion in a computer-based simulation—it could be a disease, a bit of gossip, a fact, or anything else—and had it spread in the cell phone network according to one basic assumption: The stronger the tie between two people, the more likely the contagion would spread from one person to another.

This is entirely reasonable. If you spend more time with someone who has a cold, you're more likely to get sick. The more often you speak with someone, the more likely they are to tell you a bit of juicy gossip.

For each of one thousand simulations, the team would begin by randomly choosing a few people to start the contagion. Then, at each step, a weighted coin would be flipped for each neighboring person who could possibly become infected. The stronger the tie, the more weighted the coin would be toward infection. Through running the simulations they were able to see how long it took for everyone to become infected, as well as what happened along the way.

When they tested the network and ran this experiment, they discovered that weak ties aren't that important to spreading knowledge.

While weak ties do in fact hold the network together, much as Granovetter suspected, they aren't integral for spreading facts. Weak ties, while bringing together disparate social groups, aren't strong enough to spread anything effectively.

But strong ties also aren't that important. While they can spread a fact with ease, most of the time they are spreading it to people who already know it, because strong ties only exist in highly clustered groups of people who often all know similar things.

So Granovetter wasn't quite right. Ultimately it's the medium-strength ties that are the most important. They are that happy medium between ties that are too weak to spread anything and those too strong to be found in anything but socially (and information-ally) inbred groups.

These are the types of ties that allow knowledge to spread, facts to disseminate, sometimes even errors to propagate. The people you trust a little bit but aren't your closest friends, your work friends, or something a bit more than strangers but less than a good buddy: These people provide the ties that are the most important in allowing something to spread far and wide. They connect different enough social circles that the fact can infect a new group, but they are also strong enough to provide a good chance of spreading it.

We see hints of this when it comes to spreading the printing press and Gutenberg's bundle of innovations. The key individuals in spreading new facts were other Germans in different towns. These individuals, who had the necessary knowledge and skills (and often some sort of connection to Gutenberg), were able to bring these innovations to new communities.

How facts spread and reach each of us is intimately tied to how we are connected to one another, and network science can provide us with a guide to understanding how this works. But just because a fact spreads doesn't necessarily mean that it's right. Just as quickly as truth can spread, so can facts that are wrong.

Errors are especially pernicious facts. Is there anything that science can say about how errors spread and persist in a population? To understand that we have to look at some of these errors, starting with one that I fell victim to many years ago: the celebrated case of the brontosaurus.

AS a child, I was a student of dinosaurs. When you're a six-year-old boy, this is not really a choice—it's some sort of biological imperative. I was well versed in the different types of sauropods and theropods, the dietary habits of these ancient giants, and even the recently popularized theory that they were warm-blooded rather than the slow, dim-witted, and cold-blooded dinosaurs known to previous generations.

We had a wonderful activity in my kindergarten class in which each of the eighth-grade students were paired with a kindergartner. They were assigned to interview us, to discuss our interests, hopes, and dreams, and then to write a storybook for us based on all of these findings. The subject matter for my book was never in any doubt. Dr. Sam Arbesman, paleontologist, was given the honor of being sent to South America to investigate rumors that a live dinosaur was there, and to capture it.

But despite this obsession with dinosaurs, there was one fact that I got wrong. It's entirely basic, and yet no one had told me that I was incorrect: the name brontosaurus. The four-legged saurischian, with the long neck and tiny head, is one of the iconic creatures of my youth. In fact, it was the very species of dinosaur that my fictional self was sent to South America to capture. And yet its name is actually apatosaurus.

BY the 1860s, dinosaur-fossil hunting was in full swing. Darwin had published *On the Origin of Species* the previous decade, and decades before that the argument that these fossils were the bones of creatures that perished in the deluge had been discarded. They were now bona fide monsters that had lived unbelievably long ago, and they were part of the pageantry of interconnected life explained by evolution.

Bestriding this wave of scientific discoveries were two American paleontologists: Edward Cope and Othniel Marsh. While entwined by history, they were highly unlike each other. Cope had little formal training in paleontology, and only received an honorary master's at the same time he began a position at Haverford College. Marsh, on the other hand, had a doctorate from the University of Heidelberg, was a professor at Yale University, and was the curator of the Peabody Museum of Natural History.

Beginning in 1863, Marsh and Cope began something of a competition, though that term is not quite accurate. The phrase used by some—the Great American Dinosaur Rush—also doesn't do it justice. This was no friendly rivalry over who might discover more new species. If you consider the depths of bribery, theft, and even ideological dispute to which both paleontologists sank, it is certainly more appropriate to call this conflict by its more common name: the Bone Wars.

Marsh and Cope were initially collaborators, but they soon had a falling-out. The proximate cause seems to be Marsh's pointing out an error in the reconstruction of a swimming creature known as the elasmosaurus, a sort of Loch Ness monster. This, coupled with Marsh's payment—behind Cope's back—of diggers to divert all future fossil finds in their New Jersey area to him, cemented the beginnings of the war. The conflict was exacerbated by the fact that Marsh was a staunch Darwinist, while Cope adhered to an older theory known as neo-Lamarckism. Eventually the two became fierce rivals, resorting to nearly any stratagem in order to describe more dinosaur species.

In the midst of all of this, Othniel Marsh published his discovery of the brontosaurus in 1879. Its full taxonomic name is *Brontosaurus excelsus*, essentially meaning in Latin "most sublime thunder lizard." Two years earlier, in 1877, he also submitted a paper entitled "Notice of New Dinosaurian Reptiles from the Jurassic Formation," in which he described a slightly smaller dinosaur (it was clearly a juvenile, or child, dino) that he called apatosaurus. This name means "deceptive lizard," due to Marsh's observation that its bones looked similar to those of another species. Marsh, in these descriptions, even managed to get a dig in against Cope, noting that his findings related to these species were superior to Cope's, whose "[c]onclusions based on such work will naturally be received with distrust by anatomists."

The brontosaurus discovery went on to be supplemented with a complete skeleton, beautiful to behold and the harbinger of its fame in popular culture. The apatosaurus, on the other hand, languished as a tiny collection of bones that included not much more than a pelvic bone and a shoulder blade.

Marsh and Cope continued their respective breakneck paces of dinosaur discovery, lashing out at each other over and over. Cope even went so far as to purchase a controlling interest in the distinguished scientific journal *American Naturalist* in order to make it easier for him to publish his discoveries. Despite their vitriol and animosity, they actually didn't fight any more about the brontosaurus.

But in 1903, an error was found by the paleontologist Elmer Riggs. This time it was Marsh who had gotten something wrong. While Cope didn't have the satisfaction of knowing this (both had already died several years earlier), it was a rather large error. Riggs argued that the brontosaurus was in fact simply a version of the apatosaurus. Due to the error, the brontosaurus no longer formally existed. Since it had been discovered after the apatosaurus, the apatosaurus name received precedence. And while the name brontosaurus was much more impressive, the title apatosaurus was now the correct one due to being first.

Despite this problem, many paleontologists persisted in using brontosaurus. Why confuse the public when it was a rather minor issue?

Then, in 1978, two other paleontologists, J. S. McIntosh and David Berman, noticed something even bigger that was amiss: The original brontosaurus had been graced with the wrong head! It had the head of another large, plant-eating dinosaur. They recognized that a skull, misidentified as belonging to a different species, in fact belonged to the apatosaurus. After this discovery was made they realized it was time to set this error straight; scientists began to agitate for a switch to the name apatosaurus.

But this did nothing to diminish the appeal of the name brontosaurus. This dinosaur was already out of the bag. The brontosaurus was featured in popular books of all types, including those from which I gained much of my childhood dinosaur expertise. The United States Postal Service even included it as one of four dinosaur stamps in 1989, nearly a century after the discovery of the misclassification and two decades after the beginning of the discontent in the paleontological community. I only learned of the misclassification in the early 1990s, through Stephen Jay Gould's essay "Bully for Brontosaurus," when he argued that the postal service did the right thing, even though the name was technically incorrect.

Since then, apatosaurus has been gaining currency, although rather slowly, as seen here in a Google Ngram:

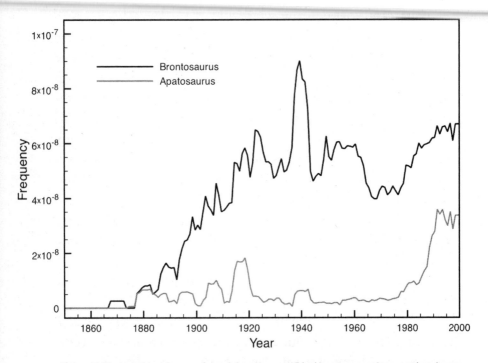

Figure 7. The number of uses of word *brontosaurus* (black) versus *apatosaurus* (gray) over time. Data courtesy of Google Books Ngrams and the Cultural Observatory.

But is this always how erroneous facts persist? Sadly, it seems that this is often the case, that there are many examples of errors that have stuck around for far longer than they should have.

IN chapter 9 we'll examine why people refuse to change their knowledge, or at least neglect to update their mental databases of facts. But how does a novel fact, even a wrong one, spread and persist in the population? Are there regularities to how errors spread?

One of the strangest examples of the spread of error is related to Popeye the Sailor. Popeye, with his odd accent and improbable forearms, used spinach to great effect, a sort of anti-Kryptonite. It gave him his strength, and perhaps his distinctive speaking style. But why did Popeye eat so much spinach? What was the reason for his obsession with such a strange food?

The truth begins more than fifty years earlier. Back in 1870, Erich von Wolf, a German chemist, examined the amount of iron within spinach, among many other green vegetables. In recording his findings, von Wolf accidentally misplaced a decimal point when transcribing data from his notebook, changing the iron content in spinach by an order of magnitude. While there are actually only 3.5 milligrams of iron in a 100-gram serving of spinach, the accepted fact became 35 milligrams. To put this in perspective, if the calculation were correct each 100-gram serving would be like eating a small piece of a paper clip.

Once this incorrect number was printed, spinach's nutritional value became legendary. So when Popeye was created, studio executives recommended he eat spinach for his strength, due to its vaunted health properties. Apparently Popeye helped increase American consumption of spinach by a third!

This error was eventually corrected in 1937, when someone rechecked the numbers. But the damage had been done. It spread and spread, and only recently has gone by the wayside, no doubt helped by Popeye's relative obscurity today. But the error was so widespread that the *British Medical Journal* published an article discussing this spinach incident in 1981, trying its best to finally debunk the issue.

Ultimately, the reason these errors spread is because it's a lot easier to spread the first thing you find, or the fact that sounds correct, than to delve deeply into the literature in search of the correct fact.

Michael Mauboussin, the chief investment strategist of Legg Mason Global Asset Management, in an article about fact-checking, relates his own experience with this sort of error propagation. While working on his book *Think Twice*, he came across an equation in a book about statistics that calculated the value of wines from the Bordeaux region. The problem was, when Mauboussin tried it, it didn't work. It turned out that his source was riddled with errors, ranging from one number that was ten times too small to another that contained a rounding error, completely changing the meaning of the equation. Only when Mauboussin tracked down the original scientific paper did he find the correct version.

There are many examples where a small error, despite being corrected later, has spread through a population. If you want to spend days poring over persistent errors that have spread far and wide, Snopes.com is a great font for these bits of information. Or even look at Wikipedia. In his delightfully nerdy Web comic *xkcd*, author Randall Munroe wishes for a world in which schoolchildren read the Wikipedia page on common misconceptions weekly, in order to learn truth as well as skepticism. Both of these sites are full of urban legends, false facts, and misconceptions that have become prevalent.

One good rule of thumb when examining how errors propagate over time is to look for a simple phrase: *contrary to popular belief*. While the phrase is a favorite of writers with a love for the counterintuitive point (and this author is not immune to this), it's also a clear indication that a bit of knowledge has spread far and wide despite being inaccurate. The antidote to this false fact, which of course the writer is about to tell you, has yet to penetrate the popular consciousness. And this phrase is by no means new. I have found instances of it in books and magazines from the nineteenth century debunking false facts about lunar phases, medical knowledge, and even the heredity of genius.

There are even examples where such misinformation has been spread purposefully, albeit sometimes with a wink rather than with malicious intent. But since we often don't track our sources, this can have a rather problematic effect. For example: I have a book on my shelf entitled *Dictionary of Theories*. Perusing it one day, I came across a curious entry:

> **Dynamics of an asteroid** (1809) *Astronomy* Initiated by C F Gauss (1777–1855), but reputed to have had its outstanding exposition in an elusive textbook by James Moriarty (c. 1840–1891), with later contributions by other mathematicians including K Weierstrass (1815–1897) and J E Littlewood (1885–1977).
>
> The motion of an asteroid, which is now generally understood as a minor planet, is that of a body of negligible gravitational attraction in the gravitational field of two massive bodies, just like a spacecraft under the influence of the Earth and the Moon.
>
> J F Bowers, "James Moriarty: A Forgotten Mathematician," New Scientist, 124 (1989). Parts 1696–7, 17–19.

While I had never heard of this theory, I wasn't terribly surprised, as the dictionary contains many obscure ideas. But I did recognize the reference, and that surprised me. That's because the citation was to a monograph by none other than James Moriarty, the arch nemesis of Sherlock Holmes.

As brilliant as he might have been, Moriarty never existed. Yet here he was, in a fictional reference that had somehow jumped into the real world. Searching a bit further, I tracked down the article referenced in *New Scientist*, and discovered that it was a tongue-in-cheek analysis, by John Bowers of the School of

Mathematics at the University of Leeds, of the mathematical contributions of Professor Moriarty, including his analysis of how gravity operates on asteroids.

But the citations to Moriarty's work didn't end there. I even found a dissertation by Kristian Kennaway, a physics doctoral student at the University of Southern California, that cites one of the other celebrated bits of research by Moriarty: a treatise on the binomial theorem, published in none other than the *Bohemian Journal of Counting*.

While this is no doubt a fun bit that Kennaway inserted to see if his committee members were paying attention (my aunt placed a banana bread recipe into her master's thesis and no one noticed), Kennaway actually cites Moriarty's mathematical work to explain something, providing the justification for a mathematical concept using the work of a fictional character.

Thus far, I haven't seen many other examples of this, and I doubt such bizarre overlaps of fiction with reality have propagated far. But it does give one pause.

Bad information can spread fast. And a first-mover advantage in information often has a pernicious effect. Whatever fact first appears in print, whether true or not, is very difficult to dislodge. Sara Lippincott, a former fact-checker for *The New Yorker*, has made this explicit. These errors "will live on and on, . . . deceiving researcher after researcher through the ages, all of whom will make new errors on the strength of the original errors, and so on into an explosion of errata." This is strong stuff. These errors become ever present and extremely difficult to correct. It's like trying to gather dandelion seeds once they have been blown to the wind.

I myself was a victim when I actually propagated the myth that a frog, if boiled slowly, will not jump out of a pot. I mentioned this in passing in the *Boston Globe*, using it to explain how people don't notice factual change if it happens slowly. I was taken to task soon after by James Fallows, of *The Atlantic*, who has worked hard to remove this falsehood from the population; in fact, the frog only remains in the pot if it's brain-dead.

Can we understand in any rigorous way how these sorts of falsehoods continue to propagate? Happily, there is scientific research that delves into how they spread. But that science requires us first to take a little detour to examine some very old typos in ancient manuscripts, their surprising relationship to genetics, and how both of these fields deal with error.

WE can look to the children's game of telephone to understand how facts can be corrupted and spread: The children sit in a circle, and one person begins by whispering a phrase or sentence to the child next to them. This person whispers to their neighbor, who in turn does the same, continuing until the person who completes the circle, the last one to hear the sentence, says aloud what they heard. This is then compared to what the first person initially said, often with hilarious results. Of course, sometimes this is because there's someone malicious somewhere along the line—the kid who delights in replacing every verb with *fart*, for example. But in general, the sentence decays without any malice or intent. It simply gets changed because hearing a whispered sentence doesn't provide great fidelity. It's what information scientists would refer to as a noisy channel. When information is passed from one person to another it has the potential to become inaccurate unless there are a whole host of error-checking mechanisms.

A clear case of this that we can actually measure and study quantitatively can be found in the world of old texts. Surprisingly, understanding the errors in these manuscripts is actually quite similar to understanding genetics. This may sound a bit odd. What do handwritten manuscripts from the medieval period or earlier have to do with genetics? On the surface, nothing: One is a distinguished part of the humanities and the other a hard experimental science. However, while those who study each of these fields have very little to do with one another, it turns out that there is a great deal of symmetry. It mainly comes down to mutation.

Scholars who study paleography—the field of research that

examines ancient writing—are all too aware of the mistakes that scribes make when copying a text. These types of errors, which can be used to understand the provenance of a document, are actually nearly identical to the types of errors caused by polymerase enzymes, the proteins responsible for copying DNA strands.

When it comes to copying DNA—those strands of information that code for proteins and so much more—there are a few advantages over simply hand copying a document. DNA's language is made up of four letters, or bases, which come in complementary pairs: A always goes with T, and G always goes with C. When DNA is copied, its double helix is unzipped, and the letters of each helix—one side of the zipper—can be easily paired with their complementary letters. This results in two new double helices—closed zippers—both of which have properly paired letters, because the complementary letters act as a simple way to prevent errors.

Nonetheless, when DNA is replicated, it's sometimes done imperfectly. The group of chemical machines responsible for duplicating a strand of DNA occasionally makes mistakes. That's what can make up a mutation: an incorrect copying, or even a piece of DNA getting hit by a cosmic ray. However it happens, some error is introduced into the sequence. For example, an A gets turned into a G, or something much bigger happens. The types of mutations fall into a few categories, such as duplicating a section of DNA or deleting a letter, due to regular ways that the DNA copying mechanisms operate. The majority of these errors cause no problem whatsoever, but in some cases a change in a single letter of DNA can cause some large-scale issues, such as in the case of sickle-cell anemia.

There are systematic errors in copying a text as well. Whether it's skipping a word or duplicating it, there is an order to the ways in which a scribe's mind wanders during his transcription. Many of the errors can be grouped into categories, just like the different types of genetic mutations. And not only are there regularities to how both DNA and ancient manuscripts are copied incorrectly, but these types of errors are often very similar, despite the large differences between how scribes and enzymes work.

There is a common scribal error known by the Greek term *homeoteleuton*. This refers to a type of deletion, in which there are two identical word phrases separated by some other text and the scribe accidentally skips to the second phrase without transcribing the intervening portion, including the first instance of the phrase. For example, there is a verse at the end of the creation story in Genesis that reads, "And on the seventh day God finished the work that he had done, and he rested on the seventh day from all the work that he had done." Notice that the phrase "the work that he had done" is repeated. If a scribe incorrectly transcribed the verse simply as "And on the seventh day God finished the work that he had done," and then proceeded to the next verse, that would be a homeoteleuton.

In genetics, this same error is known as a slipped-strand mispairing mutation. AATTCGATATACGA gets copied as AATTCGA, skipping the middle section.

Insertions can occur during copying in both genetics and paleography as well. Simply called insertions in genetics, it is called dittography for manuscripts. There are also reversals: metathesis in paleography and chromosomal transpositions in genetics. And point mutations, substituting the wrong genetic base when copying DNA, also occur in handwritten manuscripts. In both cases the wrong letter is written, based on probabilities of their being similar. In DNA, C and T are quite similar chemically and can be confused easily. In ancient Greek, lambda and delta look similar, and are more likely to be exchanged as well. And the list goes on.

While fun to chronicle such similarities, they can also be exploited in the same way. Each type of error occurs with different yet predictable frequencies, which we can use if we want to make judgments about the ages of documents or sequences. For example, if a rare mutation is found frequently in a genetic sequence, the sequence can be assumed to be quite old, since a long period of time is needed for these errors to accumulate. In addition, errors can be used to infer the relationship between differing versions of documents or sequences. If two documents have few differences between them, we can assume that they are more closely related than two documents that have many differences.

More generally, mutational differences between DNA sequences can be used to understand the evolutionary history of a population, or even of a group of species. So too with variants of the same manuscript. A famous example of this is of research that quantitatively studied the differences between the surviving versions of Geoffrey Chaucer's *The Canterbury Tales*. By subjecting the variants to a battery of genetic analyses, the researchers were able to better understand the contents of the ancestral version, Chaucer's own copy.

They used one of the better-known, and saucier, sections of Chaucer's work, "The Wife of Bath's Prologue," in order to trace how changes can be used to find the original version. Based on fifty-eight different surviving versions of this section, which is 850 lines long, the team of researchers—made up of biochemists, information scientists, and humanities scholars—used off the shelf computer programs from the field of evolutionary genetics to deduce what Chaucer's original version likely looked like. They concluded that Chaucer's original was in fact an unfinished version, complete with his notes about intended additions and deletions.

Armed with a sense of how genetic tools can be applied to understand how texts spread and change, we can now use this to understand exactly what we have been trying to grasp: how information, especially misinformation, spreads.

Mutation of texts is far from an ancient problem. It happens in modern times and we can see it especially clearly in the world of science, when facts themselves are referenced: Citations—references in scholarly papers to previous works—also mutate over time.

Too often a popular paper isn't actually read by a scientist and then cited in her own work. Sometimes scientists just look at the bibliographies of other papers and copy the citation to the paper instead. This somewhat lazy approach is unfortunately all too common, and if one scientist types it incorrectly, then suddenly there is a mutated version of the citation out in the wild. If other scientists come along and look only at that reference and not the original paper itself, that typo gets propagated from paper to paper, leading to a proliferation of errors. Just as we can learn about how ancient manuscripts spread errors, studying these mutations can allow us to learn about the history of the article that is being cited.

Mikhail Simkin and Vwani Roychowdhury, professors of electrical engineering at the University of California–Los Angeles, actually measured how often these sorts of factual corruptions occur in the scientific literature. In a series of papers, they explored the possible mechanisms for how this occurs, with most of their mathematical models relying on hypothetical scientists grabbing a few papers they've recently read and copying citations from the back. An assumption of laziness, certainly, but it also seems that something close to this might actually be the truth.

Simkin and Roychowdhury conclude, using some elegant math, that only about 20 percent of scientists who cite an article have actually read that paper. This means that four out of five scientists never take the time to track down a publication they intend to use to buttress their arguments. By examining these mutations we can trace these errors backward in time, and understand how knowledge truly spread from scientist to scientist, instead of how it appeared to spread.

We can even see the spread of such misinformation in a somewhat more lighthearted context. If you had an e-mail address in the late 1990s, you were likely the recipient of a letter that looked something like this:

```
This is for anyone who thinks NPR/PBS is
a worthwhile expenditure of $1.12/year
of their taxes. . . . A petition follows.
If you sign, please forward it on to
others. If not, please don't kill it—send
it to the e-mail address listed here:
```

> XXXX@XXXX.edu. PBS, NPR (National Public
> Radio), and the arts are facing major
> cutbacks in funding.

In case it isn't immediately clear, this is a chain letter. It is one of the less insidious types, as some are much more overtly hoaxes that promise good luck if you spread it, bad luck (or death!) if you ignore it, and the like. While certainly not an accurate piece of knowledge, these types of letters circulate for a very long time.

Once again, we can use letters to understand how errors spread by examining how they circulate before dying out. This was the question that David Liben-Nowell and Jon Kleinberg, both computer scientists, set out to answer.

Liben-Nowell and Kleinberg compiled a massive collection of different versions of the chain letter shown above, as well as a petition purporting to organize opposition to the Iraq war (neither of these letters were entirely factual and had their roots in hoaxes). Through a Web site, they asked for volunteers to search their e-mail archives for variations of them. In their database, they found all the hallmarks of biological-style textual mutation. From their paper:

> Some recipients reordered the list of names on their copy of the letter in ways closely analogous to the kinds of chromosomal rearrangements one finds due to sequence mutation events in biological settings. We observed examples of point mutations (in some petition copies, names were replaced by the names of political figures), insertion/deletion events (there were a number of small blocks of 1–5 names that were present in the middle of the list in some petition copies and absent in other copies), duplication events (blocks of 2–20 names that were duplicated in some petition copies, sometimes immediately adjacent within the list and sometimes hundreds of names later), block rearrangements (in

one petition, two pairs of blocks of 2–3 names were swapped relative to their position in all other copies that contained the same names), and one hybridization event (the names at the ends of two copies of the petition were intermingled after their common prefix in a third copy).

But just as these mutations and errors can be used to understand how knowledge spreads, another element of these letters can be used to explore the branching and spreading of information. Using the signatories (the people who spread it), which are included in the data, we can see how false knowledge makes its way through a population. By looking at how the letters accumulated signatures, the researchers were able to trace their spread and see who sent the letters to whom.

And they found something that contradicts our intuition about social networks. While we are embedded in highly clustered social circles, ones that also have the property of connecting everyone within a handful of hops (that six degrees of separation again), the spread of these chain letters does not have the feel of an epidemic. Rather than the letters spreading to hundreds of individuals, who in turn each spread it to hundreds of additional recipients themselves, and so forth, it was much more tame. They only spread successfully to one or two people at each step. So while they spread for a long time, slowly burning through a tiny sample of the population, they didn't create any sort of rapid, massive conflagration.

This can be a good thing. While a fact or, more important, an incorrect fact—whether the iron content of spinach or the proper scientific name of a long-necked dinosaur—might be able to percolate through a population, and slowly weave its way through a group, it won't necessarily spread widely. The downside, though, is that it might linger. It can hop from person to person, lasting far longer than we might expect, even if it only affects a tiny subset of the group.

Happily, it is often the case that credible information or news spreads faster and wider than what is false. But no matter the speed with which an error takes hold, rooting it out can be a very difficult process, as it's hard in everyday life to trace the error back to its source and disabuse each person at every step of their wrong information.

FACTS do not spread instantaneously, even with modern technology. They weave their way through social networks in mathematically predictable ways. Along the way they can also mutate and become filled with errors, again in a reliable manner. Errors can continue to spread, lasting much longer than we might realize. Soon enough, the knowledge in a single area is filled with facts but also with the ejecta from a single burst of errata, making it difficult to know what is true.

Luckily, there is a simple remedy: Be critical before spreading information and examine it to see what is true. Too often not knowing where one's facts came from and whether it is well-founded at all is the source of an error. We often just take things on faith.

The modern origins of empirical scientific knowledge lie in the sixteenth and seventeenth centuries. This time period, known as the Scientific Revolution, saw advances such as Newton's theory of gravitation, Boyle's gas laws, Hooke's recognition that all living things are made of cells, and the beginnings of the Royal Society—a scientific group that exists to this day. The spirit that infused this time period brought forth a whole host of new knowledge, and the disproving of facts that had existed for centuries, if not millennia. The Scientific Revolution has made the swift changes in modern-day knowledge possible.

But some of the most important components of this endeavor were to try to eliminate errors and create a means of spreading correct facts. Many of the papers presented in the early years at the Royal Society were devoted to trying to understand errors, to root out misunderstandings, or to test the veracity of tales told to them that often seemed too good to be true. For example, here is a characteristically wordy title of a paper published in 1753 in the *Philosophical Transactions* of the Royal Society: "Experimental

Examination of a White Metallic Substance Said to Be Found in the Gold Mines of the Spanish West-Indies, and There Known by the Appellations of Platina, Platina di Pinto, Juan Blanca." No doubt some man of science had heard of this mysterious white metallic substance from these gold mines and its properties (it appears to be platinum) and felt it important to examine it.

Anything that was heard they tried to test and to eliminate errors in it, however long they had persisted. Most important, they didn't keep this new knowledge secret. They spread it far and wide, publishing it and disseminating it through the loose network of natural philosophers of Europe.

One's knowledge is dependent upon it being knowable to you specifically, on it having been spread to you. As we've seen, this spread relies on social networks, and sometimes on the all-too-human tendency to corrupt information as it spreads. But as long as we remain true to the spirit of the Scientific Revolution, by not taking things on faith and by spreading true facts, we are far from being overwhelmed with error.

But sometimes, even with the massive advances in technology and our ability to disseminate knowledge—whose modern origins are found in Gutenberg's Mainz—facts sometimes don't spread as far as they should. Therein lies the curious situation of hidden knowledge.

CHAPTER 6
Hidden Knowledge

MY father, Harvey Arbesman, is a dermatologist and an epidemiologist. He spends about half of his time seeing patients, diagnosing and treating skin cancer, and the other half doing research. As a researcher, he is fond of the unexpected hypothesis and the counterintuitive concept. This has led him to publish on such topics as whether malignant melanomas are associated with the increased use of antibacterial soaps and whether dairy consumption is related to acne. Essentially, he is drawn to the tough challenge. This research style led him to InnoCentive.

Alpheus Bingham was the vice president of research and development strategy at the pharmaceutical company Eli Lilly when he began thinking about experts and how they solve problems. He realized that while an expert might solve a hard problem 20 percent of the time, simply giving it to five experts won't always yield results. There's a good chance that all the experts will fail.

But what if this pool of people was made much wider? Perhaps, Bingham argued, there was a "long tail of expertise" (his term, not mine) of lots of people who are all interested in solving a technical problem but each of whom has a very small chance of success. Using this logic, as long as you get a really large group there's a decent chance that the problem will be solved. The math sounds like it should work out, but would it really work in practice?

Bingham, with the support of Eli Lilly, created a separate company called InnoCentive, which is designed to test this hypothesis. InnoCentive acts as a clearinghouse between organizations or companies that have problems and solvers—those people from all areas of life who are interested in solving problems and can work better in the aggregate than the experts.

Bingham's intuition was right: InnoCentive works. It works because it draws on solutions and insights from different fields. Often the solver is involved in a technical discipline that is near the area of the problem but just different enough to be distinct. With this distinctiveness comes the potential for informational import and export. A fact or solution might be well-known in one area, but it is still an entirely open question in the other. This allows people who might not be experts to bring what they know in their field and apply it to other areas. A sort of fact recombination—where ideas are brought together in new ways—is often the way that problems are solved at InnoCentive.

For example, when Roche brought a problem that it had been working on for fifteen years, the crowd recapitulated all the possible solutions that the company had already tried, and in only sixty days. But even better, there was an actual working solution among the proposals, something the company had failed to find. When NASA used InnoCentive, they quickly got the answer to a problem that had been bothering them for thirty years! Instead of working in the same area for a quarter of a century, you can open up the question to a larger group and get an answer from an unexpected source. And more important, an unexpected field.

So my father took a look at InnoCentive to see if it had any interesting problems that he could try to solve. While it was originally populated by engineers, chemists, and other technically minded individuals, it is rapidly expanding and broadening its focus, including into the life sciences, so my father felt at home on the Web site.

While sifting through InnoCentive's e-mail challenges, my father came across one that intrigued him. The Prize4Life Foundation, a nonprofit organization devoted to curing and treating amyotrophic lateral sclerosis (ALS), also known as Lou Gehrig's disease, was offering a prize for a stepping-stone toward an eventual cure.

Curing ALS is hard. Rather than tackling the problem wholesale, since the final solution is essentially unknowable, Prize4Life broke down its goals into smaller tasks. One of the first major problems they wanted to solve was the creation of something known as a biomarker, a way of measuring the progression of the disease. For many diseases, even those for which there is no known cure, there are clear ways of measuring how far along the disease is within an individual. For example, in HIV research, one biomarker used is the level of cytokines, small molecules found in the immune system. However, there weren't any known biomarkers for ALS based on chemicals or anything internal to the patient. The only way to determine the progression of this disease used some established ways of measuring what the patient was still able to do, and these correlated with how long the patient had left to live.

Prize4Life's InnoCentive challenge sought promising hypotheses for potential, noninvasive biomarkers, with a prize of $15,000. The organization also created another challenge, which offered a prize of $1 million if the effectiveness of a biomarker could actually be demonstrated through testing in patients.

My father is not a neurologist, nor does he have any specialized knowledge in the area of neurodegenerative diseases. However, he is an expert in something else: undiscovered public knowledge.

In the mid-eighties, a professor of library science named Don Swanson realized that, for all of our presumed prowess at organizing knowledge, we were falling profoundly short. While we had made great strides from the days of Linnaeus and his taxonomy of living things, as massive amounts of information have become digitized, we were becoming confronted with a filtering problem.

Remember, this was the middle of the 1980s. The first graphical Web browser was nearly a decade away. But Swanson presciently realized that sifting through information successfully is a far from trivial task, and even if we navigate all that we find, there was still much knowledge that was being ignored.

While technological knowledge constantly increases, as shown in chapter 4, in other areas knowledge sometimes churns, as when scientific facts are overturned. Sometimes what we know can suffer a regress, such as when the library of Alexandria was destroyed. Similarly, there is often knowledge that is in fact hidden in plain sight, yet to be discovered and used. These findings, which Swanson termed *undiscovered public knowledge*, are extreme special cases of those examples from chapter 5, in which knowledge is not spread far and wide.

What does it mean for knowledge to exist but remain hidden? It's one thing for a result to be ignored—that's a specific instance of this sort of knowledge, which will be discussed later. But Swanson was talking about research that, due to its inability to combine with other findings, is less valuable than it could be. Imagine that in one area of the scientific literature there was a paper showing that A implies B. Then, somewhere else, in some seldom-read journal (or even a journal read only by those in an entirely different area), an article contained the finding that B implies C. But since no one has read both papers, the obvious result—that A implies C—remained dormant, hidden in the literature as an unknown fact.

Could this really happen? Swanson argued that not only was it possible but that there are many such examples. He demonstrated this with a novel finding: Swanson combined research that showed that dietary fish oil improves blood circulation with entirely separate research that associated poor blood circulation with a condition known as Raynaud's syndrome. Aside from a few scattered reports, no one, until Swanson, had combined the findings of both these areas and recognized the mechanism that could allow fish oil to help those with Raynaud's syndrome. Swanson, far from a biologist, was even able to publish this finding in a medical journal.

Building on this, Swanson continued to develop methods of combing through the newly digitized literature. He expanded his use of MEDLINE, an online database run by the National Library of Medicine, which is housed at the National Institutes of Health. MEDLINE allowed Swanson to search rapidly for medical key words, and then to combine research that had remained separate. In the late 1980s, such databases were still in their relative infancy, but Swanson recognized their potential.

In 1996, a decade after his initial paper, Swanson and a collaborator, Neil Smalheiser, revisited undiscovered public knowledge to see if it was simply a matter of one exceptional example or if it had become a generalizable concept with broad applications, through the use of computational resources. They highlighted six additional examples in this follow-up that revealed intriguing connections between things such as magnesium deficiency and migraines, and sleep habits and phospholipases—a type of enzyme. But while digitizing information is powerful and could reveal a handful of examples, this was but the tip of the iceberg.

MY father had studied information science and the idea of undiscovered public knowledge. So he did what a dermatologist, untrained in neurology but mindful of the concept of undiscovered public knowledge, would do: He looked in the neurological literature for a hint of some dermatological relationship to ALS that had remained hidden. After combing through many papers, he was led to a book from 1880, where the first case of ALS was described. The physician author noted that patients with this newly described condition did not develop bedsores, even as the disease caused them to be entirely immobile and bedridden.

This old book in turn led my father to reports by neurologists who had used biopsies of ALS patients to examine the structure of the skin's framework. They found that there were specific chemical differences related to a decrease in the stretchiness of an ALS patient's skin, known as its elasticity. These scientists had found that, even though this was a neurological disease, it could affect other parts of the body, including the skin. My father realized that this was the critical clue.

My father knew that in dermatology there are numerous skin diseases that also reduced skin elasticity. But unlike in ALS, there were well-understood quantitative methods of measuring the changes in elasticity due to these diseases. He connected these two areas that had been entirely unrelated and proposed that the techniques for measuring skin elasticity could be used as a way of measuring the biochemical changes in ALS patients, but in an entirely noninvasive manner. After writing this theory up, my father submitted it to InnoCentive.

And he won, as one of five hypotheses selected.

Prize money in hand, my father went to researchers at the dermatology and neurology departments of Columbia University and suggested that they all collaborate on testing his hypothesis. They conducted a pilot study with ALS patients, and it seems that my father's undiscovered public knowledge was right: There are quantitative changes as the patients got sicker.

While my father didn't win the final prize—that went to a team from Harvard—his team eventually won a monetary prize honoring how much progress they had made. Furthermore, some neurologists are currently exploring the connections between ALS and the skin. My father also showed me in a tangible way the power of connecting pieces of hidden knowledge.

Hidden knowledge takes many forms. At its most basic level hidden knowledge can consist of pieces of information that are unknown, or are known only to a few, and, for all practical purposes, still need to be revealed. Other times hidden knowledge includes facts that are part of undiscovered public knowledge, when bits of knowledge need to be connected to other pieces of information in order to yield new facts. Knowledge can be hidden in all sorts of ways, and new facts can only be created if this knowledge is recognized and exploited.

But, happily, hidden knowledge—and fact excavation—isn't simply a matter of reading lots of papers and hoping for the best.

There is a science to understanding how new facts can be uncovered in what is already known.

ONE of the most fundamental rules of hidden knowledge is the lesson learned from InnoCentive: a long tail of expertise—everyday people in large numbers—has a greater chance of solving a problem than do the experts. The problems that go on to be solved by InnoCentive are precisely the ones that experts can't solve. This was true even before computers and large scientific databases. Revealing hidden knowledge through the power of the crowd has been a great idea at almost any time in human history.

This is behind the concept of the innovation prize. The British government once offered a prize for the first solution to accurately measure longitude at sea—created in 1714 and awarded to John Harrison in 1773—but this was by no means the first such prize. Other governments had previously offered other prizes for longitude: the Netherlands in 1627 and Spain as early as 1567. They hoped that by getting enough people to work on this problem, the solution—perhaps obtained by drawing on ideas from different fields—would emerge.

In 1771, a French academy offered a prize for finding a vegetable that would provide adequate nutrition during a time of famine. The prize was won two years later by Antoine Parmentier for his suggestion of the potato. To our ears, this sounds so obvious as to be silly. But at the time, the potato, due to its origins in South America, was generally unknown in France. And those who did know of it thought it was involved in outbreaks of leprosy. Parmentier's research into starch, and his willingness to look into unplumbed areas of knowledge, provided the opportunity to uncover a fact—that potatoes are nutritious, not deadly—that was known in other parts of the world but had remained hidden in Europe.

But while prizes help tease out innovations and ideas that would otherwise remain hidden and accelerate the pace of knowledge diffusion, hidden facts unfortunately remain a far too common part of how knowledge works.

IN 1999, Albert-László Barabási and Réka Albert wrote a celebrated paper that was published in *Science*, one of the world's premier scientific journals, about a process they termed *preferential attachment*. The process is responsible for creating a certain pattern of connections in networks—also known as a long tail of popularity—by the simple rule of the rich getting richer, or in this case, connections begetting more connections. For example, on Twitter there are a few individuals with millions of followers, while most users have only a handful. This paper shows how, by assuming a simple rule that newcomers look at everyone in the network and are more likely to connect with the most popular people, you can explain why you get the properties of the entire network—in Twitter or elsewhere—that we see. Using a wide variety of datasets and some mathematics, they showed this rigorous result.

Unfortunately, they weren't the first. Derek Price, the father of scientometrics, had written a paper in the 1970s showing that one can get this same pattern by invoking a similar rule with respect to how scientific papers cite one another. But Barabási and Albert didn't know about Price.

Price wasn't the first either. Herbert Simon, a renowned economist, had developed the same idea in the fifties. Which also happened to be the same concept that Udny Yule had published several decades earlier.

The general concept of preferential attachment is actually known by many names. It's known as the Matthew effect, as Robert Merton coined it, in sociology, and is related to something known as Gibrat's Law when it comes to understanding how cities and firms grow.

More generally, Mikhail Simkin and Vwani Roychowdhury, the same scientists who explored the errors in scientific citations, examined a few models developed in physics that are widely used to explain certain types of probability distributions. This includes the models behind everything from chain reactions to income distributions. They explored how these models have been reinvented again and again, and they go into great detail, over the course of thirty-five pages, eventually summarizing these successive reinventions in a large table. They show, for example, that something known as the branching process was discovered in the mid-1840s, only to be rediscovered in the 1870s, then again in 1922, 1930, 1938, 1941, and 1944. The Erdős-Rényi random graph, written about by Paul Erdős and Alfréd Rényi in 1960, was first examined in 1941 by Paul Flory, the chemist and Nobel laureate. As Stigler's Law of Eponymy states: "No scientific law is named after its discoverer." Naturally, Stephen Stigler attributes this law to Robert Merton.

Extreme cases of this can especially be found during times of war. Richard Feynman, the celebrated physicist, shared the Nobel Prize with another physicist, Sin-Itiro Tomonaga. Tomonaga had come to the same scientific results independently of Feynman, working in Japan during World War II without the benefit of interacting with the scientific world of the West.

This sort of situation was most pronounced during the many decades of the Cold War. During the latter half of the twentieth century scientists in the West often duplicated research done in the Soviet Union. Certain concepts in computer science, related to the difficulty of mathematical problems, were independently discovered both in the United States and Soviet Union. In addition, a precursor to the laser was independently developed both in the East and the West. In a widespread way, knowledge was effectively hidden from a large portion of the world, and duplication was the natural result.

Far from duplication of discovery being a strange, isolated situation common only during times of war (or at least cold war), this seems to occur quite often. Known as *multiple independent discovery*, some have occurred five or more times simultaneously and can make innovation seem nearly inevitable. Classic examples of simultaneous innovation are the telephone, for which two patents were filed on the same day, the discovery of helium, and even the theory of natural selection, which was proposed by both Charles Darwin and Alfred Russel Wallace.

In some of these cases (though by no means all), there was a certain amount of delay: A discovery was simply not known by one party and ended up being duplicated, sometimes years later. If knowledge had spread widely, such a thing would not have occurred. But knowledge can be hidden for other reasons. There are occasions in science when knowledge is hidden because it is so far ahead of its time.

THERE is a rule in online circles known as Godwin's law. It states that as the length of an Internet discussion approaches infinity, the probability that someone will be compared to Hitler or the Nazis approaches one. Perhaps a corollary should be that if a discussion is about Shakespeare, the longer it is, the likelihood of arguing that Shakespeare did not actually write the plays attributed to him increases.

I'm not going to weigh in on the question of whether William Shakespeare authored the plays attributed to William Shakespeare. However, the concern that leads to this discussion—the comparison of Shakespeare's background and training to what he actually produced—is not unique to literature. In fact, there are many Shakespeares, in all fields, including mathematics and science. There are individuals who, based on their environment, seem highly unlikely to have done what they did. Their accomplishments give us a sense of how knowledge can move forward. And one of these Shakespeares was George Green.

George Green was a miller who lived in the town of Nottingham during the nineteenth century. He received one year of schooling when he was eight years old, in 1801, and then went to work in the mill and bakery of his father. Until nearly the end of his short life— he died at the age of forty-seven—he was completely unknown. Yet this entirely unremarkable man, whose background consisted of knowledge related to grain and baked goods, produced a variety of unbelievable innovations in mathematics and physics. Two of his contributions are known as Green's theorem and Green's functions, the latter of which are complex enough to vex many mathematicians and physicists.

His first work, *An Essay on the Application of Mathematical Analysis to the Theories of Electricity and Magnetism*, introduced both of these concepts, and it was published in 1828. Frankly, no one is entirely sure what happened between 1801 and 1828, and what led to Green laying the foundations for the mathematics eventually used in quantum mechanics. There was one other person trained in mathematics in Nottingham, and while we don't know if Green and this man interacted, they did live near each other. And frankly, this seems to be the only possible solution to this mystery. But however Green learned advanced mathematics, Einstein once remarked that Green's contributions were decades ahead of what was expected. As a result of his work being so ahead of its time, and coming from far outside the mainstream, Green's work was almost completely unknown until after his death in 1841.

But Green, the Shakespeare of mathematical physics, is not the only example of these sorts of people. There are many instances when knowledge is not recognized or not combined, because it's created by people who are simply too far ahead of their time, or who come from backgrounds that are so different from what is traditionally expected for scientific insight. For example, Gregor Mendel, now recognized as the father of genetics, died without being known at all. It wasn't until years after his death that the Augustinian monk's work was rediscovered, due to other scientists doing similar experiments and stumbling upon his findings. Yet he laid the foundation for the concepts of genes and the mathematics of the heritability of discrete traits.

There is also Charles Babbage, who designed the first mechanism for a programmable computer, but who had the misfortune of living during the Industrial Revolution, when he was unable to construct his invention, mainly because it was far too expensive at the time. His Difference Engine No. 2 actually had parts that corresponded exactly to the memory and processors found in modern computers.

So facts can remain hidden for a long time, whether the problem is because they are very advanced or because they come from a different discipline. But is there a way to measure this? Specifically, how often is knowledge skipped over?

A recent study in the *Annals of Internal Medicine* examined this phenomenon in a quantitative and rigorous fashion. Karen Robinson and Steven Goodman, located at the Johns Hopkins University, wanted to see how often scientists were aware of previous research before they conducted a clinical trial. If science properly grows by accreting information, it should take into account everything that has come before it. But do scientists actually do that? Based on everything I've mentioned so far, the answer likely will be no. But then, what fraction of the time do we ignore (or simply don't know about) what has come before us?

Robinson and Goodman set out to see how often scientists who perform a clinical trial in a specific field cite the relevant literature when publishing their results. For example, if a clinical trial related to heart attack treatment is performed, Robinson and Goodman wanted to see how many of these trials cite the trials in that area that had come before it. While a clinical trial needn't cite every paper that preceded it, it should provide an overview of the relevant literature. But how to decide which papers are relevant and which ones aren't? Rather than be accused of subjectivity, or have to gain expertise in countless specific areas, Robinson and Goodman sidestepped these problems by doing something clever: They looked at meta-analyses.

Meta-analysis is a well-known technique that can be used to extract more meaning from specific papers than could be gained from looking at each one alone. A meta-analysis combines the results of papers in a specific area in order to see if there is a consensus or if more precise results can be found. They are like the analyses of thermal conductivity for different elements mentioned in chapter 3, which use the results from lots of different articles to get a better picture of the shape of what we know about how these elements conduct heat.

Assuming the meta-analyses bring together all the relevant trials, Robinson and Goodman simply looked through all the studies examined in each meta-analysis to see how many of the studies cited in the meta-analyses were also mentioned each of the newer studies being examined.

What they found shouldn't be surprising. Scientists cite fewer than 25 percent of the relevant trials when writing about their own research. The more papers in the field, the smaller the fraction of previous papers that were quoted in a new study. Astonishingly, no matter how many trials had been done before in that area, half the time only two or fewer studies were cited.

Not only are a small fraction of the relevant studies being cited, there's a systematic bias: The newer ones are far more likely to be mentioned. This shouldn't be surprising after our discussion of citation decay and obsolescence in chapter 3. And it is hardly surprising that scientists might use the literature quite selectively, perhaps to bolster their own research. But when it comes to papers that are current, relevant, and necessary for the complete picture of the current state of a scientific question, this is unfortunate.

Imagine if we actually combined all the knowledge in a single field, and if scientists actually read all the analyses that their work was based on. What would happen to facts then? Would it make any difference?

Quite a bit, it turns out.

IN 1992, a team of scientists from the hospitals and schools associated with Harvard University performed a new type of analysis. These researchers, Joseph Lau and his colleagues, examined all the previously published randomized clinical trials related to heart attacks. Specifically, they looked at all trials that were related to the use of a drug called a streptokinase to treat these heart attacks. Combing through the literature, they found that there were thirty-three trials between the years 1959 and 1988 that used this treatment and examined its effectiveness.

Why did they stop at 1988 instead of going all the way up to 1992? Because 1988 was the year that a very large study was published, finally showing definitively that intravenous streptokinase helped to treat heart attacks. But Lau and his colleagues did something clever.

Lau lined up the trials chronologically and examined each of their findings, one after the other. The team discovered something intriguing. Imagine you have just completed a clinical trial with your drug treatment of choice. But instead of just analyzing the results of your own trial, you combine your data with that of all of the studies previously completed up until then, making the dataset larger and richer. If you did that, Lau and his colleagues discovered, a researcher would have known that intravenous streptokinase was an effective treatment years before this finding was actually published. According to their research, scientists could have found a statistically significant result in 1973, rather than in 1988, and after only eight trials, if they had combined the disparate facts.

This type of analysis is known as *cumulative meta-analysis*. What Lau and his colleagues realized was that meta-analyses can be viewed as a ratchet rather than simply an aggregation process, with each study moving scientific knowledge a little closer to the truth. This is ultimately what science should be: an accumulation of bits of knowledge, moving ever forward, or at least sweeping away error as best we can. Lau and his colleagues simply recognized that to be serious about this idea of cumulative knowledge, you have to truly combine all that we know and see what new facts we can learn.

While Don Swanson combined papers from scientific areas that should have overlapped but didn't, Lau and his colleagues combined papers from very similar areas that had never been combined, looking at them more carefully than they had been examined up until then. By using cumulative meta-analysis, hidden knowledge could have been revealed fifteen years earlier than it actually was and helped improve the health of countless individuals.

Modern technology is beginning to aid cumulative meta-analysis and its development, and we can even now use computational techniques to employ Swanson's methods on a grand scale.

WE are not yet at the stage where we can loose computers upon the stores of human knowledge only to return a week later with discoveries that would supplant those of Einstein or Newton in our scientific pantheon. But computational methods are helpful. Working in concert with people—we are still needed to sort the wheat from the chaff—these programs can connect scientific areas that ought to be speaking to one another yet haven't. These automatic techniques help to stitch together different fields until the interconnectivity between the different areas becomes clear.

In the fall of 2010, a team of scientists in the Netherlands published the first results of a project called CoPub Discovery. Their previous work had involved the creation of a massive database based on the co-occurrence of words in articles. If two papers both have the terms *p53* and *oncogenesis*, for example, they would be linked more strongly than words with no two key terms in common. CoPub Discovery involved creating a new program that mines their database for unknown relationships between genes and diseases.

Essentially, CoPub Discovery automates the method that Don Swanson used to detect the relationship between fish oil and Raynaud's syndrome but on a much larger scale. It can detect relationships between thousands of genes and thousands of diseases, gene pathways, and even the effectiveness of different drugs. Doing this automatically allows many possible discoveries to be detected. In addition, CoPub Discovery also has a careful system of checks designed to sift out false positives—instances where the program might say there is an association when there really isn't.

And it works! The program was able to find a number of exciting new associations between genes and the diseases that they may cause, ones that had never before been written about in the literature.

For example, there is a condition known as Graves' disease that normally causes hyperthyroidism, a condition in which the thyroid produces too much hormone. Symptoms include heat intolerance and eyes that stick out more prominently, yielding a somewhat bug-eyed appearance for sufferers. CoPub Discovery, when automatically plowing through the large database, found a number of genes that had never before been implicated in Graves' that might be involved in causing the disease. Specifically, it found a large cluster of genes related to something known as *programmed cell death.*

Programmed cell death is not nearly as scary as it sounds. Our bodies often require the death of individual cells in order to perform correctly, and there is a set of genes in our cells tailored for this purpose. For example, during embryonic development, our hands initially have webbing between the fingers. But prior to birth the cells in the webbing are given the signal to die, causing us to not have webbed hands. Webbed hands and feet only occur when the signal is given incorrectly, or when these genes don't work properly.

What CoPub Discovery computationally hypothesized is that when these programmed cell death genes don't work properly in other ways, a cascade of effects might follow, eventually leading to the condition known as Graves' disease. CoPub Discovery has also found relationships between drugs and diseases and determined other previously unknown effects of currently used drugs. For example, while a medicine might be used to help treatment for a specific condition, not all of its effects might be known. Using the CoPub Discovery engine and the concept of undiscovered public knowledge, it becomes possible to actually see what the other effects of such a drug might be.

The researchers behind CoPub Discovery did something even more impressive. Rather than simply put forth a tool and a number of computationally generated hypotheses—although this is impressive by itself—they actually tested some of the discoveries in the laboratory. They wanted to see if these pieces of newly revealed knowledge are actually true. Specifically, CoPub Discovery predicted that two drugs, dephostatin and damnacanthal, could be used to slow the reproduction and proliferation of a group of cells. They found that the drugs actually worked—the larger the dose of these drugs, the more the cells' growth was inhibited. This concept is known as drug repurposing, where hidden knowledge is used to determine that medicines are useful in treating conditions or diseases entirely different from their original purposes. One of the most celebrated examples of drug repurposing is Viagra, which was originally designed to treat hypertension. While Viagra proved effective for that condition, many of the participants in the clinical trials reported a certain intriguing side effect, also making Viagra trials one of the only cases where the pills left over at the end of the study were not returned by the participants.

There are many other examples of computational discovery that combine multiple pieces of knowledge to reach novel conclusions. From software designed to find undiscovered patterns in the patent literature to the numerous computerized systems devoted to drug repurposing, this approach is growing rapidly. In fact, within mathematics, there is even a whole field of automated theorem proving. Armed with nothing but various axioms and theorems well-known to the mathematics community, as well as a set of rules for how to logically infer one thing from another, a computer simply goes about combining axioms and other theorems in order to prove new ones.

Given enough computational power, these systems can yield quite novel results. Of course, most of the output is rather simple and pedestrian, but they can generate new and interesting provably true mathematical statements as well. One of the earliest examples of these is Automated Mathematician, created by Doug Lenat in the 1970s. This program constructed regularities and equalities, with Lenat even claiming that the Automated Mathematician rediscovered a fundamental unsolved problem (though, sadly, did not solve it) in abstract number theory known as Goldbach's Conjecture. Goldbach's Conjecture is the elegant hypothesis that every even number greater than two can be expressed as the sum of two prime numbers. For example, 8 is 5 + 3 and 18 is 7 + 11. This type of program has provided a foundation for other automated proof systems, such as TheoryMine, briefly mentioned in chapter 2, which names a novel, computationally created and proved theorem after oneself or a friend, for a small price.

TheoryMine was created by a group of researchers in the School of Informatics at the University of Edinburgh. While some people might be excited to simply have something named after themselves and ignore the details, TheoryMine will give you not only the theorem but also a capsule summary of how the theorem was proven. The theorems are all related to the properties of functions and for most people are rather opaque. Nonetheless, it's great that a mechanism to discover a piece of hidden knowledge is available for a consumer audience.

In addition to automated discoveries, there are now even automated scientists, software capable of detecting regularities in data and making more abstract discoveries. A computer program known as Eureqa was developed by Mike Schmidt, a graduate student at Cornell University (and current president of Nutonian, Inc.), and Hod Lipson, a professor at Cornell. It does something a bit different from the other projects mentioned already: Given lots of data, it attempts to find meaning in an otherwise meaningless jumble of facts.

Eureqa takes in a vast quantity of data points. Let's say you're studying a bridge and trying to understand why it wobbles. Or an ecosystem, and how the relative amounts of predators and prey change over time. You dump all the data you've collected into Eureqa—how many predators there are on each day, as well as the quantity of prey, for example—and it attempts to find meaning.

Eureqa does this by using a simple technique known as *evolutionary programming*; due to its computing power, this technique is very powerful. Eureqa randomly generates a large variety of equations that could conceivably explain relationships between the changes in data. For example, it will create an equation that attempts to mathematically combine the inputs of your system and show how they can yield the outputs. Of course, if it's given a random equation, the odds are very good that it will have absolutely no insight into the underlying phenomena it is trying to explain. Instead of explaining the data, it will spit out gibberish.

But what is randomly generated doesn't have to be satisfactory. Instead, a population of random equations can be evolved. Just as biological evolution can result in a solution—an organism that is well adapted to its environment—the same thing can be done with digital organisms. In this case, the equations are allowed to reproduce, mutate, swap bits of their formulas, and more. And this is all in the service of explaining the data set. The better the equations adhere to the data, the more they are allowed to reproduce.

Doing this over and over results in a population of good and fit equations, formulas that are far cries from the initial, randomly generated ones. Eureqa can even yield equations that can actually generate findings as complex as the concept of the conservation of energy, one of the foundations of thermodynamics.

In the case of these automated-discovery programs, the more knowledge we have available, the more raw materials we have for these programs. The more data, the more new facts these programs can in turn reveal.

So it's important for us to understand how knowledge is maintained if we want to make sure we can have the maximum amount of data for these automated-discovery programs. Specifically, is most knowledge actually preserved? Or are the raw materials for hidden knowledge that we have available only a remnant of what we might truly know?

THE Middle Ages, far from being the Dark Ages, as some of us might have been taught, was a time of science and innovation. Europeans developed medical techniques and made advances in such areas as wind energy and gunpowder.

But it was also a time for preservation. That many of the texts written in ancient times, or even in the early Middle Ages, would make it to the modern era was by no means a foregone conclusion.

As mentioned in the previous chapter, prior to the printing press manuscripts had to be copied by hand in order for information to spread.

I have a book on my shelf entitled *The Book of Lost Books: An Incomplete History of All the Great Books You'll Never Read*. It's a discussion by Stuart Kelly of books that have been lost to time, whose names we know or from which excerpts have been passed down, but whose full texts are unknown. There is a long history of such references, even going back to *The Book of the Wars of the Lord*, a lost book that the Bible itself references when quoting a short description of the location of ancient tribal boundaries in Numbers 21:14–15.

The Book of Lost Books is organized by author, and the names of those whose books we don't have is astonishing: Alexander Pope, Gottfried Leibniz, William Shakespeare, Charles Dickens, Franz Kafka, Edward Gibbon, and many more. And, of course, there are many ancient and medieval writers. From Ovid and Menander to Ahmad ad-Daqiqi and Widsith the Wide-Traveled, we know of many writers whose works have not been preserved. And then there is the Venerable Bede.

The Venerable Bede was a Christian scholar and monk from England in the late seventh and early eighth centuries. In addition to being a very holy man in the eyes of the Catholic Church—he was made a doctor of the Church in 1899, a title indicating a person's importance on theological and doctrinal thought, as well as being given his Venerable title only a few decades after his death, and later canonized—he was also a man of history and science. In fact, he has become known as the father of English history due to his *History of the English Church and People*. He also wrote books about mathematics, such as *De temporum ratione*, which discusses how to quickly perform mathematical calculations by hand.

But for all that we have of Bede's work, even more of it no longer exists. Bede wrote a great deal of English poetry, nearly all of which has been lost. However, we have most of the Venerable Bede's important scholarly works. In order for knowledge to be discovered, and to be recombined in novel ways for new facts to be unearthed, it first needs to be preserved. For every one of Bede's books, how many books of others do we not even have a memory? How much knowledge is fated to remain forever hidden?

A Cornell professor of earth and atmospheric sciences named John Cisne decided to tackle this question. Using the works of the Venerable Bede as a guide, he employed the same technique I did in chapter 5: He brought biology to the world of handwritten manuscripts. While I examined mutations and evolution, Cisne wanted to see how often medieval manuscripts go extinct while being copied. So he used population and demographic models to understand how Bede's works spread.

Just as organisms reproduce and die out, so do manuscripts, in a fashion: They "reproduce" by being copied and can die by being lost or destroyed. Reproduction in this case follows a logistic curve, just like bacteria in a petri dish. Since books cannot grow without bound, a logistic curve is a more realistic function to use than a simple exponential. The logistic allows for growth in which there is a certain carrying capacity—the maximum number of copies of a book that can be made.

Cisne used mathematics from the world of population biology to describe this simple state of affairs, and he was able to create a model that fit the number of the Venerable Bede's technical books that have survived from each century. From this Cisne arrived at the likelihood that a document would be copied. Specifically, he found that documents from the Middle Ages were fifteen to thirty times more likely to be copied than destroyed.

In addition, he calculated the half-lives of these books: how long they would last before the destruction of half of the copies. He found that the half-lives were between four and nine centuries, a surprisingly long time. Cisne was able to conclude that most documents from the early Middle Ages, and perhaps even antiquity, have, in fact, survived.

We have certainly lost a great deal; time ravages much knowledge. But when it comes to hidden knowledge, it's heartening to know that many facts aren't lost to history; they can indeed be discovered.

SO we now know that facts are seldom lost. And as long as knowledge is preserved, we have the raw materials for unearthing hidden knowledge. As we've seen earlier in the chapter, that still doesn't prevent much of knowledge from remaining hidden—witness everything from Mendelian genetics to the true import of clinical trials—but through modern technology, we now have computational ways of connecting and recombining disparate bits of knowledge to create new facts.

In fact, hidden knowledge and its discovery is no longer the domain of the medieval scholar or information scientist, or even of the robotic mathematician. Tools related to hidden knowledge are being created for everyone, enabling a certain renaissance in the discovery of knowledge; facts can be spread and mixed in novel ways, unburied and shown the sunlight. One of these tools is Mendeley, which is designed for the average scientist.

One of the most annoying and tedious parts of publishing scientific work is in the details—specifically, the details of formatting. Each journal has its own specific rules for fonts, the organization of the paper's content, and, most maddening, how to format the citations. When you write a paper, you carefully format each reference specifically for the journal to which you are submitting. But woe betide the scientist whose paper is rejected, forcing her to reformat for another journal and a resubmission. And lest you be surprised by this tedium, multiple submissions are more often the rule than the exception.

Into this detail-oriented morass have stepped a number of computational tools to help deal with these issues. The most popular of these is EndNote. This is a computer program that allows references to be imported easily from scientific databases, or to be entered manually. Creating a bibliography for a specific journal becomes as simple as selecting it from a drop-down menu. Want to submit to *Nature*? Easy. Rejected from *Nature* and now aiming a paper at *Proceedings of the National Academy of Sciences*? This too is a simple matter, requiring little more than a single click.

But a new online tool has arrived recently, called Mendeley. In addition to simplifying reference importation, synchronizing one's bibliography online, and many other wonderful features, it has another: social networking. Instead of the scientist simply working with the set of references they use to write their papers in isolation, it allows them to see their friends' references; it acts as a sort of social network for scientists.

As Mendeley grows in popularity—and it seems that it's hitting the critical mass that's necessary for any social Web site to thrive—it allows for the collaborative exposure of knowledge that each of us individually hasn't been aware of.

But it provides another important feature: It allows scientists to see articles that are related to ones that they're already looking at. By automatically finding topic relationships between papers, Mendeley brings undiscovered public knowledge to the scientific masses. Scientists can now find a paper on psychology that can shed light onto network science or a math paper that can help with X-ray crystallography. In doing so it can help create new facts.

These capabilities are even being brought to the everyday user. There are a wide variety of computer tools that allow someone to collect snippets of information—quotations, references, pictures, articles, Web pages, and more—in a simple and searchable place. Some store these notes in the cloud, some on a desktop, and some even allow these notes to be shared with others.

However, one program has an ability that others lack: DEVONthink uses something called *semantic and associative data processing*. The other programs require searching for certain words or combinations of words. If a note has these words, it shows up; otherwise the program can't find it. DEVONthink, however, is a bit more clever. It uses a special computational technique to analyze the entire text and find relationships between words. So if the search is for the word *house* and there is a note that is a quotation about the wonderful nature of the home, DEVONthink is likely to find such a relationship. It can also tell which notes are similar to one another, providing cognitive connections that are not always available to us, since we can't hold thousands, or even hundreds, of notes in our minds at once. But computers can, and they can draw the connections for us, providing the substrate for new facts and bits of knowledge.

Steven Johnson, a writer whose books rely on the connectedness of disparate ideas, uses DEVONthink a great deal, and he reports that it has benefited him greatly. From an essay in which he praises this tool's powers, he gives an example:

> This can create almost lyrical connections between ideas. I'm now working on a project that involves the history of the London sewers. The other day I ran a search that included the word "sewage" several times. Because the software knows the word "waste" is often used alongside "sewage" it directed me to a quote that explained the way bones evolved in vertebrate bodies: by repurposing the calcium waste products created by the metabolism of cells.
>
> That might seem like an errant result, but it sent me off on a long and fruitful tangent into the way complex systems—whether cities or bodies—find productive uses for the waste they create.

New facts are all around us. And due to the algorithmic properties of modern technology, we now have the possibility of discovering them.

DIGGING up hidden knowledge is now far from an impossibility, or even from being solely the domain of the specialist; it has become eminently possible and easy. Knowledge doesn't get lost or destroyed any longer, and that seems to have happened even less often than we used to believe. Facts are now commonly digitized, and are ripe for being combined and turned into new facts. We are in a golden age of revealing hidden knowledge.

When this happens it can sometimes lead to drastic, sudden changes in what we know. These changes—when what we know is fundamentally overhauled in an abrupt and dramatic way—are also subject to quantitative regularity, and are the subject of the next chapter.

CHAPTER 7
Fact Phase Transitions

IN 1750, Thomas Wright, a British astronomer, published a diagram in his book *An Original Theory or New Hypothesis of the Universe*. This diagram showed a whole host of stars. But there was more in the diagram than just stars; the stars weren't alone. Each star was surrounded by a small cloud of orbits, an entire planetary system. What Wright was clearly implying was that our sun was not particularly special: Other planets orbited around every star, much like the ones in our solar system.

At that point in history this notion was nothing more than a hope, and a somewhat sacrilegious one at that. It was no more than a logical deduction derived from the Copernican notion that our place in the universe need not be a particularly privileged one.

More recently, in Carl Sagan's 1980 documentary *Cosmos*, Sagan is shown speaking to a classroom of schoolchildren. In his characteristically excited and inspirational manner, he speaks of our solar system and hands out pictures of the different planets and moons to each of the students. Then he begins to muse about ideas that are a bit more speculative but which are just as exciting. Explaining the fundamentals of detecting extrasolar planets— planets outside the solar system—he tells them that humanity will discover such planets in their lifetimes. He predicts that we would one day find other planets like our own Earth, as well as ones similar to all the other planets throughout our solar system.

We have had this yearning for centuries, to know of the existence of worlds orbiting other stars like ours. It would give us a sense of our place in the universe and flesh out the true details of our stellar neighborhood. While to some these discoveries would make our stellar home a bit more ordinary (and a few find it a very worrisome idea), others have been sure that this feeling of drabness would certainly be offset by how these planets outside our solar system can provide us a way of viewing ourselves.

When Sagan spoke to those schoolchildren, he was right. In 1995, the hypothetical became the factual when the team of Michel Mayor and Didier Queloz announced the discovery of a planet orbiting 51 Pegasi, a star much like our own sun.

Since 1995, thousands of such exoplanet candidates have been detected by various methods. These planets vary widely in their characteristics, with many far larger than Jupiter and orbiting closer than Mercury. But the first discovery was something special. In addition to a higher likelihood of having its place in the history books, 51 Pegasi produced a rapid shift in our knowledge. Humanity, in the course of a single issue of *Nature*, overhauled its view of the universe. We went from knowing of no planets orbiting sun-like stars like our own to knowing that they exist. To oversimplify, everything before that discovery was speculation, and everything else after that was simply collecting more examples of the same: more extrasolar planets.

WHEN facts change we can often anticipate the speed at which the change occurs. Populations grow according to certain rules, medical knowledge accumulates in a regular fashion, and new technologies allow us to do things at faster and faster rates—but all in a way that is well understood and regular.

However, there are other facts that don't seem to adhere to this sort of logic. Knowing DNA's structure, or whether Pluto was a planet, or that airplanes were possible—all of these happened in extremely rapid shifts. The iPhone appeared so rapidly in the world of technology that executives from a rival company thought many of its claimed specifications were lies, and Marc Andreessen has argued that it's as if it appeared from the future, incredibly ahead of its time. One day there was a certain understanding of how we thought the world works, and the next day humanity's factual environment had undergone a fundamental change.

But can these actually be explained? Astonishingly enough, there is in fact an order to these rapid shifts in our knowledge. We can find regularities in them, and sometimes even predict them before they happen.

This type of rapid change in knowledge—when we go from one state of awareness to another—is one example of a larger class of phenomena in science that are termed *phase transitions*. This term is well-known in physics, and most of us are no doubt familiar with them on a daily basis. When water freezes, when dry ice becomes carbon dioxide (by a process known as sublimation), even when gold is melted—all of these are examples of matter changing its phase. These are so much a part of our lives that we do not marvel when they happen. But while everyday occurrences, phase transitions are intriguing to physicists for a simple and powerful reason: They are clear cases when small changes make a big difference.

In general, through a small change of an underlying parameter, such as temperature, we get a small change in the overall properties of what we're looking at. Warm a cup of water by a small amount and it becomes a bit hotter. Or put metal in a furnace and it becomes warm to the touch.

But at some magical point a tiny shift in the underlying parameter induces a rapid and pervasive change in the system. Warm that water just a little bit more, and suddenly it's not just a warmer liquid—it's a gas: steam.

While entirely unextraordinary, something complicated is happening at the microscopic level that leads to these changes.

What is it about the boiling point of water that allows a small change in temperature to produce a massive change in the overall structure of all of its molecules? Or, in a slightly less familiar area, why does heating a magnet cause it to lose its magnetization (and even weirder, when it's cooled, to stay demagnetized)? In the parlance of condensed matter physics, the branch of physics that examines these phenomena: What causes this cascading behavior and resulting phase transition?

And what about facts? Can the same sort of rapid cascade occur in the world of facts, when a large-scale shift in what we know about the world is due to some smaller underlying change?

The answers to these questions can be found in a simple mathematical model of how magnets work, developed by a physicist named Ernst Ising.

THERE are all sorts of mathematical models for physical systems. Some try to actually mimic the complex mess we see around us, the most well-known of these being weather models. We don't just want to understand how weather changes; we want to know how likely it is that it will rain tomorrow. So we input temperature details from throughout the country; wind speeds; barometric pressure values measured over time; and much more. These values are then put into complex equations and simulated in a powerful computer, allowing us to see what they all will be in the future, inside the computational world that has been created.

But these kinds of models, while very powerful, don't allow us to say anything general about them. We can't write down a simple equation for how the average air temperature of the entire country will affect rain, because the model is far too complex for that. To understand that sort of thing, or any other system for which we want to explain a certain phenomenon, we need to create much simpler models. These don't make any claims for verisimilitude. Instead, they go to the other extreme and claim the following: We can make an extremely basic model that even with all the complexity of real life stripped away still has certain features of our complicated world. And if we can capture these features of our world, maybe we can understand why they occur. In our case, the question is whether a simple model can be made that exhibits phase transitions. This is exactly what Ising set out to do.

How does the Ising model work? Imagine we lay out a large deck of cards, in which each card is black on one side and white on the other, into a grid. When the cards are all on the same side—either black or white—the system is considered to be in one state, like a solid. Entirely uniform and understandable. However, if the cards are flipped randomly, and there are no regions of a single color, we have something much more irregular. We know such a system as a gas.

How does this system change? We choose a card and flip it. If we start with all the cards on their black sides, pretty soon we'll start getting cards showing white, and soon enough we will have something that looks random and fits what I described as a gas.

As I've explained the system thus far, it seems like we're just flipping cards at random. And a moment's thought shows that flipping the cards at random will result in no overall change. If any part of a random grid of black and white cards is changed at random, the details of the picture will change: which specific cards are black and which are white. But if you zoom out, we still get the same overall picture: static.

But here's where the model gets interesting. If, whenever we flip a card, we have the possibility of also flipping its neighboring cards to the same color, this little twist is enough to yield strikingly different behavior. In the Ising model, whether an individual card is black or white depends on two things: the "temperature" of the system and the neighbors.

The parameter that we call *temperature* is related to how likely it is that a card gets flipped when its neighbor is flipped. At high temperatures lots of cards flip, but simply because it's hot and the temperature is making everything jumpy. But as the system cools down, things start becoming more clumpy, because now neighboring cards are affected when a card is flipped. So whole patches of cards can change color quite rapidly.

When the temperature drops low enough, the grid of cards will snap into a single color, either white or black. Now we're in an entirely different state, which we can call a solid, since it's a solid block of color.

Here's what Ising recognized: There is a temperature value that can be solved for mathematically and that precisely represents the transition between the two phases: between when everything is the same color and when it's all random.

The Ising model—which is referenced by thousands of articles in fields that range from biology to the social sciences—is a mathematical abstraction that has been used to explain all sorts of phase transitions, many outside the realm of physics. Phase transitions have been used to understand all types of rapid change, from ecological models that explore ecosystem collapse and abrupt climate change to tipping points for how fads and fashions spread.

But what about facts? Can these sorts of models provide insight into how changes in knowledge operate?

It turns out that knowledge is not that different from abstract magnetic models. While the underlying changes in our knowledge develop over time, we can have big sudden jumps in something else we might be examining. In the Ising model, we change something slowly and continuously—temperature—and this change yields a sudden jump in something else: the state of the system, from gas to solid, for example. In the world of facts, this sort of thing can also happen: A sudden jump in our knowledge might be due to some other facts changing and being slowly accumulated. But this still sounds very abstract; how might this work in practice?

ONE of the biggest facts in human history was overturned abruptly, and it was done in a single footprint: Since the dawn of history, no human had walked on the moon. That was a fact whose status

hadn't budged at all for thousands of years. At no point did someone walk on the moon even a little bit. At 10:56 P.M. EDT on July
20, 1969, this all changed with a single step.

But could this phase transition have been predicted as growing
out of some underlying regular change? Yes. There were underlying
quantifiable patterns occurring over the previous decade that
would have allowed the prediction of this phase transition in humanity's place in the solar system.

Prior to Apollo 11, multiple unmanned and manned Apollo
missions had been launched. Apollo 10 actually did everything but
land on the Moon—it left the earth's orbit, circled the moon, and
returned home. It even came within fifteen kilometers of the moon's
surface. Despite what appears to be an abrupt transition in what
we know, there was a steady progression of underlying changes
that can explain the jump.

In fact, it's not even the few missions just prior to the moon
landing that could help explain the transition; hundreds of years of
smooth data can actually be used to explain this steady march. In
1953, Air Force scientists in the Office of Scientific Research and
Development created a simple chart of the fastest man-made vehi-
cles at each moment in history for the previous couple of centuries.
They found something incredible: If they extrapolated the exponential curve outward, the data showed that speeds necessary to
leave the Earth's surface could be reached within four years! A
curve had implied the existence of the first artificial satellite well
before Sputnik's launch, and exactly as predicted (Sputnik went
into orbit on October 4, 1957).

Furthermore, the Air Force scientists realized that the chart
predicted that the speeds required to land on the moon were reachable a few years after that. And the moon landing came to pass,
also right on schedule. Interstellar spacecraft are also included in
the plot, and while we haven't succeeded in that domain yet, NASA
recently announced an initiative to begin thinking about ships with
the capability of reaching the stars.

What this means is that the factual phase transition of the

moon landing was long in the cards—it was simply due to a steady change in the underlying speeds we could achieve.

These sorts of phase transitions occur in many places in the world of facts. As mentioned before, Pluto was a planet for nearly a hundred years, and then, suddenly, it wasn't. Everyone knew that smoking was fine for you. Doctors even endorsed certain brands. Then, in the 1950s, it became abundantly clear that smoking kills.

But these can likely be explained by underlying steadier changes. The detection of larger and larger objects at the fringe of the solar system meant that Pluto's status was due for reconsideration. And the proliferation of medical studies meant that it was only a matter of time before we would learn of the dangers of smoking.

In contrast, the number of mammals that we know of is a slower, steadier progression of knowledge. No single discovery greatly changes our awareness of the number of species. If we know of thousands of mammals, no single additional discovery is going to overturn that fact.

Sometimes, though, the line between these two types of knowledge change is a bit less clear. For example, while finding the first extrasolar planet is a phase transition, going from 400 to 401 known extrasolar planets is a more subtle and steady progression. In these cases, and many others like them, knowledge accretes slowly, and often according to the regular patterns discussed in chapter 2.

But understanding these steady progressions is the key to understanding the rapid shifts in our knowledge. These steady progressions are the underlying slow temperature changes that result in the fast phase transition that we see when looking at everything from a different scale. Of course, it's not always that easy. Figuring out the right underlying change to measure requires a bit of ingenuity. But doing this can help explain the many rapid shifts in our knowledge that are all around us.

. . .

EVER since the first planet was discovered outside our solar system in 1995, anticipated by Carl Sagan in *Cosmos*, we have actually been waiting for another large-scale shift in our awareness of our place in the universe. We have been searching for a planet that is truly like Earth, or one that could at least potentially be habitable. As one of the leaders in the field told me, this is ultimately what many of the astronomers in the area of extrasolar planetary research are after. Of course, there are many other topics to study, such as the true distribution of the types and sizes of planets that exist in our stellar neighborhood. But a paramount goal has been to find a planet truly similar to our own.

Finding one that actually harbors life will be a huge milestone for humanity, and it will be one of those massive phase transitions in how we view the world around us.

But just like landing on the moon, a path to such a discovery proceeds by increments. While we are fairly certain that life can exist in far stranger spaces than we can possibly imagine, good initial spaces to look are planets that are similar to Earth. And short of being able to examine the atmospheres of these planets for such hallmarks of life as oxygen, we use simpler proxies. These proxies take the form of mass and temperature. If a planet can have something like liquid water due to its surface temperature, and is about the same size as Earth (meaning our style of life could have evolved there), this planet is deemed potentially habitable, at least as a first pass.

So while detecting the first potentially habitable planet outside our solar system is a stunningly discontinuous jump in our knowledge, can it be predicted? This is what Gregory Laughlin, an astronomer at the University of California, Santa Cruz, and I tried to do in the summer of 2010. Greg is an expert on extrasolar planets, has discovered numerous ones himself, and writes the premier blog for the field, *Systemic* at oklo.org.

Greg and I knew that the discovery of a roughly Earth-like

planet was on the horizon. NASA's Kepler mission had been running for more than a year and, in fact, in mid–June 2010 had released a whole slew of exoplanet candidates, tantalizingly withholding the most promising candidates until the following February. We knew that we were in a special window of time, both for the discovery itself and for making a prediction.

We created a simple metric of habitability. Each previously discovered planet was rated on a scale from 0 to 1, where 0 is not habitable at all and 1 is quite like Earth, at least when it comes to mass and temperature. Most planets are 0 or very close to 0, since most planets are either scorchingly hot or unbelievably cold (and sometimes both at different parts of their year). But when we looked at the highest habitability values over time, we saw a clear upward trend: By focusing on the most Earth-like planet discovered each year and charting their habitability values, we found that these have been steadily proceeding upward over time. In fact, this upward trend conformed to our old friend the logistic curve. Just as scientific output in general fits exponential and logistic curves, the march toward the discovery of a potentially Earth-like planet fits one of these omnipresent functions as well.

Since it conformed to a mathematical shape, we could predict when these values would reach something that was habitable by extrapolating these curves into the future. Essentially, the properties of the discovered planets and their values of habitability act as the underlying temperature of the system. As time proceeds, the highest habitability values of discovered planets steadily increase. When we reach a very high habitable value, we jump in our knowledge: We go from not knowing of any planets that are potentially like Earth to knowing of one. Extrasolar planet discovery data allow us to see the microscopic behavior, the small underlying changes in what we know. And checking the news for the discovery of an Earth-like planet provides a metric for whether a phase transition in our knowledge has actually occurred.

After running such an analysis ten thousand times (in order to make sure our calculations were robust), we found that there was a two-thirds chance that an Earth-like planet would be discovered by the end of 2013. But even more exciting, there was a 50 percent chance of such a discovery happening by early to mid-2011!

Soon after our paper was accepted, on September 1, 2010, we published it online. Four weeks later, on September 29, 2010, a team from UC Santa Cruz and the Carnegie Institution for Science announced the discovery of the first planet truly similar to Earth—one that could actually sustain life—called Gliese 581g.

This planet has some curious properties: For example, one side always faces its star and one side always faces toward the night sky. Due to this, only in the region of the planet in permanent dusk is the temperature potentially habitable. And barely; it rates a balmy Moscow winter. But in addition to having a dimmer red sun than our own (much like the sun of Superman's home planet of Krypton), it has a most intriguing property: It might not exist.

It turns out that there is debate over whether its signature is authentic or simply a whole lot of really exciting noise. When another team examined a subset of the data, they found no evidence of Gliese 581g. Now it might just be such a subtle signal that it requires a lot of data to detect its presence. It might also not be there. However, since then, a somewhat potentially Earth-like planet, Kepler 22b, has been discovered orbiting a star about six hundred light-years away.

I have heard an astronomer involved in exoplanet discovery mention that nothing would make him happier than to become a biologist once the target planet has been discovered. And it seems that we now have crossed a certain threshold in our scientific knowledge. Where there were once only hints of such a planet, gleaned from bits of data and inferences, we now seem to be at the dawn of a new type of awareness. And this phase transition occurred due to the regular underlying variation in habitability, which was deduced from discoveries of the masses and temperatures of planets.

SOMETIMES, though, it's not that easy to determine the underlying change and see when a phase transition will happen. Even so, there are still mathematical regularities to how sudden changes in our knowledge occur, at least in the aggregate. One of these areas is in the proof of mathematical concepts.

Back at the end of 2010 I attempted to use mathematical modeling to determine when a famous unsolved problem in mathematics would be proved. This was a lot harder than the planetary discovery area. There wasn't some fundamental underlying property that was slowly changing, so that when all the individual increments were added up we would get something qualitatively new, like a planet that is potentially like Earth. Instead, when it comes to proofs, it often seems that it's just a lot of mathematicians being wrong, year after year, with little hope of finding the correct solution.

In fact, it's not quite like that. Even in failure there can be success when trying to solve something, although it's not exactly what we might have hoped for. For example, Fermat's Last Theorem was a famously long-unsolved problem. This idea was created by the Frenchman Pierre de Fermat, a lawyer by profession and mathematician by hobby, in 1637. Fermat wrote that there are no three positive numbers a, b, or c that can satisfy the equation $a^n + b^n = c^n$, if n is greater than 2. If n is 2, we get the Pythagorean Theorem, which has tons of solutions. But make n larger, and Fermat stated that no number would work in the equation. Fermat didn't prove this. Instead, in one of the most maddening episodes in math history, he scribbled this idea in the margins of a book and wrote that he had a brilliant proof, but, alas, the margin was too small to contain it.

We now think he might have been mistaken. But no one had found any numbers greater than 2 that fit the equation since he wrote this statement. So it was assumed to be true, but no one could prove it. This elegant problem in number theory had gone unproven since the seventeenth century, until Andrew Wiles completed a proof in 1995, using pages and pages of very complex math, which would most certainly not have fit in Fermat's margin. But, crucially, along the way, mathematicians proved other, smaller, proofs in their quest to crack Fermat's Last Theorem. When finally solved, whole new pieces of math were involved in the construction of the proof.

I decided to tackle predicting the proof of one of the most famous unsolved problems, something known as P versus NP. It's so tough and important that the Clay Mathematics Institute has offered a prize of $1 million for its solution. It essentially asks whether two classes of mathematical problems are equivalent. P problems are easy, and can be solved in a straightforward fashion by today's computers, and sometimes even by pencil and paper. NP problems are problems in which we can check whether an answer is correct very easily, independent of whether they seem easy to solve. Specifically, some NP problems appear to be incredibly difficult, but vexingly, if I were to give you the right answer to any NP problem, you would know right away whether it was a correct solution.

For example, finding the prime factors of a number is an NP problem. Given a large enough number (like something that's about four hundred digits long), we currently think that we would have to use all the computational resources on the planet to try to factor it. And even then it would take far longer than the lifetime of the Earth to determine what two prime numbers were multiplied to get the four-hundred-digit number.

But if I gave you the correct factors, you could simply multiply them together and see if it yielded the really big solution. The question that people have been trying to solve for the past few decades is whether NP problems—which are easy to check, such as factoring primes—are also easy to solve. Thus far, no one has found any shortcuts to solving many fiendishly difficult NP problems, and they seem to be separate from P problems, making us think that P does not equal NP. But we could just be missing some basic insight or mathematical trick that would make solving these problems much easier, and which would imply that P equals NP. For example, trying to see if a big number is divisible by three might be hard, unless you know the trick: If the sum of a number's digits is divisible by three, then so is the number itself.

Assuming that NP problems are very hard is actually the basic assumption behind modern cryptography. Codes are only strong so long as P emphatically does not equal NP. And many other things remain hard to solve if P is not equal to NP, from handling resources and logistics in large organizations, to managing complex financial portfolios, to dealing with games and puzzles such as crosswords and Minesweeper. But we don't know if they're equal or not. And we haven't known for over forty years.

When will we know? Possibly never. There are some problems that can never be solved, and in fact have been proven unsolvable. But there are many problems that have gone unsolved for hundreds of years, and then one day a mathematician puts the finishing touches on the proof. Therein lies the key for tackling this problem. I realized that I could look at the distribution of the amount of time it has taken for lots and lots of well-known unsolved problems to be solved eventually and compare that to the amount of time that P versus NP has been an outstanding problem. If P versus NP is not special but is similar to many other famous problems that have been hard to decipher, perhaps then I could gain some insight into looking statistically at how long it takes for other problems to eventually be solved.

So while I can't know exactly when any individual problem will be solved in advance, I can look at the aggregate of hard problems and see if there are any regularities. In doing this we can begin to put some bounds on our uncertainty about solving hard problems.

It turns out that if we do this, we get some very interesting distributions, known as heavy-tailed distributions. This means that there is no single amount of time we can expect to wait for a solution: Some famous problems go decades before being solved, and some, those that exist far out in the tail of the distribution, remain outstanding for hundreds of years. There was even a famous conjecture (or unproven statement) in the data set that took more than fifteen hundred years before it was eventually proven.

But using this distribution, we can see the number of years it takes for half of all of these problems to be solved, taking that 50 percent mark and using it as a likely timeframe for its solution. So given that it is soluble, it turns out we will have to wait until 2024, when P versus NP turns fifty-three, for a solution, assuming it behaves like other long-unsolved problems.

There's still a lot of uncertainty in this year. But through probability, we can now understand how rapid changes in mathematical knowledge can occur, at least in the aggregate. Given that we eventually do discover whether P equals NP, we can be certain that this change in knowledge will occur rapidly, descending upon the greater world without much warning.

Mathematics—from Ising models to probability—can help us to understand how rapid changes in the facts we know can occur around us. But are these phase transitions the rule or the exception? What should we expect more of: slow and steady changes in knowledge or extremely rapid shifts in the facts around us?

While there will no doubt always be slow change in knowledge, many of us have an intuitive sense that facts are changing around us faster and faster, with rapid transitions occurring more often every day. There is scientific evidence to buttress this intuition. To understand this we have to understand how cities produce innovation.

RECENTLY, physicists have begun to take mathematical tools from their own field and apply them to understanding the relationship between the populations of cities and how they use energy and produce new ideas. Specifically, Luís Bettencourt and his colleagues, who are affiliated with the Santa Fe Institute, found that there are economies of scale for certain properties of cities. For example, the larger the population of a city, the smaller the number of gas stations that are necessary per capita; gas stations might be indicative of energy usage of the city as a whole, and it seems that larger cities are more efficient consumers of energy. This is similar to how larger organisms are more energy efficient than smaller ones.

However, when looking at productivity and innovation, cities obey mathematical relationships that operate like increasing returns.

For example, the yearly number of patents produced in a city per person is higher for bigger cities, and in a mathematically precise way. This sort of scaling is called superlinear, because things grow faster than they would at linear speeds, faster than a straight line. Double the population of a city, and it doesn't simply double its productivity; it yields productivity and innovation that is more than doubled. These relationships have been found in patents, a city's gross metropolitan product, research and development budgets, and even the presence of so-called supercreative individuals, such as artists and academics.

When resource availability and consumption adhere to sublinear scaling and drives growth, the way it does for living things, a system develops according to a simple progression: Start small and grow rapidly, but eventually slow growth until a mature adult size is reached. However, if a system's growth is dependent upon superlinear phenomena, as in the case of cities and innovation, the mathematics require the system to grow faster and faster, until it approaches an infinite growth rate.

But infinite growth can't happen, either for organisms (cancer eventually overwhelms its own resources) or for cities. For cities, then, the only way to not be overwhelmed by this infinite growth is to undergo what the researchers deemed paradigmatic innovations, essentially to reset the parameters of growth. These innovations can encompass such changes as modern sewage systems—abundant waste limits the sizes of cities—or the birth of the skyscraper, which allows for increased urban population and density through the use of the third dimension. Whatever they are, these innovations allow the city to avoid becoming overwhelmed by its own growth. In past years these innovative resets occurred every few hundred years, and for any single person, these large-scale changes in facts were manageable. But no longer; we are living during the first time in history when multiple rapid changes can occur within a single human lifetime.

As knowledge changes more and more rapidly, the resulting change in society can be drastic. Rather than changes in degree, we have changes in kind. For someone living in a small English village at any point during the early Middle Ages, aside from certain details, there would be little difference in one's lifestyle between any two years. There might be alterations in fashion, but in one's overall life—manner of occupation, means of cooking and doing household chores—none of these would be different. In fact, even fashion during the Middle Ages only changed about every fifty years, far slower than shifting every decade or so, which has been happening since the nineteenth century in industrialized countries. Even if you lived during a rare innovative reset during the later medieval period (such as during the introduction of gunpowder), things weren't so difficult to adapt to, as they occurred only once every several generations.

This has been true for most of human history. But with exponential increases in technology and innovation, these changes are coming much more rapidly. When changes are able to occur very quickly, we are in a special situation: The world around us seems to be ever poised on the edge of some rapid shift in facts and knowledge. A small change can cause a large shift in our knowledge at any moment.

Of course, the world of facts is not the only system that can be in this sort of state; there are many other systems that can have this property. Imagine a complex ecological system in which the slightest change, such as the removal of a single species or the introduction of a pathogen, has the potential to upset the entire system. Or a party, where one person leaving suddenly kills the entire gathering. Or, at an even more basic level, imagine a pile of sand. Take a pile of sand and add a few grains, and the pile gets bigger. But as sand is added a bit more at a time, eventually adding just a few more grains triggers a rapid shift, a sort of avalanche of sand. Why the transition? What about that single grain yields a system that is right on the edge of a rapid shift?

This question was examined in great detail by three physicists: Per Bak, Chao Tang, and Kurt Wiesenfeld. In 1987, they published a simple mathematical model that aimed to understand why small changes can yield a system that is always on the verge of large shifts.

This model uses a grid, just like the Ising model. But here, each location is a spot where a grain of sand can be added. And, according to mathematical rules, the model dictates what should happen if the pile of grains at a single location gets too high. Essentially, if the height goes above a single specified value, the grains begin moving to neighboring points, similar to how water will run down a cone if poured onto its tip. When this simulated sandpile reaches what is known as a critical state, it exists in a situation exactly as described above: With the addition of each additional grain, there is absolutely no telling what will happen. There could be a tiny little movement of the sand, or a massive avalanche could be unleashed. Like Jenga or Ker Plunk, but with more mathematics, the system is constantly hovering at the brink of the unknown.

It turns out that this sort of system, one that organizes itself to always be at the edge of a total avalanche, is the hallmark of actual systems we find in the real world, perhaps even including the world of knowledge. While a bit overly metaphorical, a world with the constant potential for rapid knowledge change would look just like ours.

WE are alive during an amazing time, one in which the potential for rapid changes in knowledge is all around us. Every day that we read the news we have the possibility of being confronted with a fact about our world that is wildly different from what we thought we knew. From Pluto no longer being a planet to humans walking on the moon—to limit our examples to outer space—that is exactly what has become the norm in our modern world.

But it turns out that these rapid changes, while true phase transitions in our knowledge, are not unexpected or random. We understand how they behave in the aggregate, through the use of probability, but we can also predict these changes by searching for the slower, regular changes in our knowledge that underlie them.

Fast changes in facts, just like everything else we've seen, have an order to them. One that is measurable and predictable.

But what about measurement itself? We've explored one fact after another, but they can only really exist if we are able to quantify them. How measurement affects what we know is the subject of the next chapter.

CHAPTER 8
Mount Everest and the Discovery of Error

IN 1800, the British Empire conceived of the Great Trigonometrical Survey, also known as the Survey of India. After the British took control of the Indian subcontinent, they were in need of accurate and detailed maps. How could you properly rule a subcontinent if you didn't really know what it looked like? So Colonel William Lambton, the surveyor general of India, began this massive project of determining the precise locations of places throughout the colony.

Starting at the southernmost tip of India in 1808, Lambton employed the triangulation technique. Using trigonometry, chains, metallic bars, monuments, theodolites—a type of optical measuring device—and observations of the stars, one can in general measure the distances and locations of nearly anything. Far from being sorcery, though the varied tools employed could lead one to that conclusion, this was a well-established means of surveying a land.

The empire gradually came to understand India more and more precisely over the course of many decades. This massive project was a multigenerational one, a sort of scientific equivalent to the construction of the pyramids. When Lambton died the project was placed into the capable hands of a scientist named George Everest. Everest was in turn replaced by Andrew Scott Waugh in 1843. And under Waugh, the advanced survey techniques of the Great Arc of India, another name for this massive project, finally reached the Himalayas.

Waugh knew that the mountains in that region were some of the tallest in the world, but he did not know their height. Measurements were made over several years, and then computed. It is unknown which individual was responsible for the calculation that a peak known as XV was the tallest in the world. But in 1856, Waugh announced that they had found the world's tallest mountain, and he named it after his predecessor, Everest.

WE know that Mount Everest was determined to be the tallest mountain in the world, and that fact doesn't seem to be changing, but what exactly is its height? By 1954, the observations that had been made varied by seventeen feet. Since then, though, the variation in calculations has decreased greatly. This has been due to innovations in measurement techniques. Now, rather than having to compare measurements between multiple survey stations perched atop surrounding peaks, surveyors minimize differences by using the Global Positioning System.

In fact, the increase in precision has actually allowed us to know a new type of fact: The height of the mountain actually changes a bit every year. Mount Everest's height seems to be subject to two competing forces. On the one hand, the collision of two continental plates (Asia and India) causes a certain amount of uplift each year, perhaps about a centimeter or so, although there seems to be some disagreement. On the other hand, other forces, such as erosion and melting glaciers, can cause a decrease in height. While it's unclear how much it changes each year, we now know for certain that the height is never exactly constant. We also know that Mount Everest is moving laterally at quite a nice clip: six centimeters per year, making its location also a mesofact, one of those slowly changing pieces of knowledge. But only due to improvements in how we measure the world could we have learned these things.

Consider the world's tallest tree. By far the tallest trees in the world are the redwoods of California, coming in at nearly four hundred feet tall. The next tallest tree species are a type of eucalyptus in southern Australia, which top out around three hundred feet.

While it's possible to measure the height of any tree to within a few feet, using a laser range finder, when it comes to finding the world record holder, a laser, unfortunately, isn't precise enough. Instead, the tree actually has to be climbed in order to be properly measured.

Through a combination of measurement, discovery, and lots of tree climbing on the part of botanists and surveyors, as of 2006, the world record for the tallest tree has hopped from Rockefeller Tree (356 feet) to the Libbey Tree (367.8 feet) to the Stratosphere Giant (368.6 feet) to Hyperion (379.1 feet). While the overall record holder likely won't change appreciably in height from year to year, many trees, just like Mount Everest, also undergo changes in their height. But it's a different type of change in this case, though often just as predictable: growth. But only once we know the true height can we measure how it grows.

Ultimately, what facts are and how they change often comes down to measurement. Just as there have been systematic and quantifiable rules that govern many of the facts in our lives, measurement itself also obeys mathematics. Our increases in measurement follow certain well-defined regularities. In chapter 4, I showed how the description of our world using scientific prefixes has proceeded according to exponential growth. While we generally think of this in terms of technological growth—we start saying *gigabyte* instead of *megabyte* when our computers become more powerful—it can also be used to define our measurements, and our uncertainty. We now think of things in terms of billionths of centimeters instead of tenths of centimeters, because we have better rulers and measuring devices.

Inscribed on the University of Chicago's Social Science Research Building is a saying by Lord Kelvin: "When you cannot measure, your knowledge is meager and unsatisfactory." While this can be viewed within the context of exploring the merits of quantitative analysis as compared to qualitative examination, it also can help us think about how we measure more precisely.

Sinan Aral, a professor at the New York University Stern School of Business, has stated: "Revolutions in science have often been preceded by revolutions in measurement." From the incorrect number of chromosomes to the misclassification of species, our increased preoccupation with measuring our surroundings allows us to both increase our knowledge and find opportunities in which large amounts of our knowledge will be overturned. A corollary to Lord Kelvin's adage: If you can measure it, it can also be measured incorrectly. Measurement affects nearly everything we know, and this chapter is devoted to the many ways that measurement is intertwined with facts, beginning with how we have improved our measurement and analysis over time, and in turn our understanding of the world. Let's start with the Scientific Revolution, when many minds were preoccupied with measurement.

CHRISTOPHER Wren, known today primarily for his architecture and work in the rebuilding of London after the Great Fire in 1666, was involved in many aspects of the creation of the modern scientific endeavor. Among these were his innovations in measurement. Along with John Wilkins, another important figure of the Scientific Revolution mentioned earlier, Wren was involved in the creation of the concept of the meter.

In April 1668 Wilkins proposed at a meeting of the Royal Society that it was time for measurement to be standardized. Among those measurement standards he proposed was that of length. He argued that the base unit of length, from which all others would be derived, should be known as the *standard* and should be defined as follows: the length of a pendulum that causes it to swing from one side to the other—known as a single half-period—every second. This definition used the insight gained by Galileo decades earlier that all pendulums of equal length—regardless of the weights at the end of the pendulums—swing at the same rate. Furthermore, no matter at what height you let go of the pendulum, it takes the same amount of time to go from one side to the other.

This suggestion, which was made to Wilkins by Christopher Wren, yields the definition of a standard as thirty-nine and a quarter inches, remarkably similar to the current measure of a meter. Wilkins went on to define a regular system of lengths derived from the standard, such as a tenth of a standard being denoted a *foot*, and ten standards equaling a *pearch*. A cube with sides of a standard was proposed to be equal to a *bushel*.

It's probably clear that these derived measurements didn't stick. In fact, I would be hard-pressed to find anyone who knows the term *pearch* (in case you're wondering, it's similar to our decameter, which seems to be similarly unused). However, the standard, which was transmuted into the French *metre*, or *meter*, continues to exist today.

But in the eighteenth century another approach to defining our units of length was the method that eventually won out. Rather than using time to calculate a meter—which Wilkins argued was uniform throughout the universe and would therefore be hard to beat for constructing a unit of measurement—the other approach derived the meter from the distance between the equator and the North Pole. A meter then becomes one ten-millionth of this distance. Due to the variation in gravity over the surface of the Earth, which would affect a pendulum's swing, the French Academy of Sciences chose the distance-based measurement in 1791.

But there was a hitch. In addition to no one having actually yet visited the North Pole in 1791, the measurement methods used to calculate the distance from there to the equator were of varying qualities. Unlike in the previous cases discussed, not only were the properties of the Earth not completely known, but these unknown properties have a curious effect on the very measurements themselves. Since the imprecision of measuring the world affects the units that are being measured in the first place, there is a certain circularity when it comes to measurement. This creates a feedback loop in which the better we know how to measure various quantities, the more we improve the very nature of measurement itself.

The story of the meter has been one of ever-changing definition. Over time, the definition of the meter has evolved, as technologies have advanced and as different techniques have been proposed. As the meter's definition has changed, its precision has increased, which ultimately is the point of any effective definition.

While knowing the approximate length of a meter is helpful for many tasks, such as cutting a carpet or measuring one's own height, it will not do when it comes to finer and more precise tasks, such as designing a circuit board. As the world's complexity has progressed alongside technological and scientific development, more detailed and more exact measurements have become necessary. While I don't particularly care if my height is off by a half centimeter or so, when it comes to measuring the size of microscopic organisms, I'm going to be a bit more punctilious.

So, in 1889, an actual metal bar, made of iridium and platinum, was constructed to be the official meter and to avoid the ambiguity of the distance to the North Pole. All measuring sticks would then be based on the dimensions of this literally quintessential meter.

But that still didn't suffice, for this bar could still undergo deterioration. In addition, any slight change in the atmosphere or temperature can change its size, albeit very slightly. These considerations were included in the definition by specifying the pressure and temperature of the bar's environment, but such precise conditions are very difficult to maintain.

An international group of scientists then constructed more fundamental definitions, first using the wavelength of the emission of a certain isotope of the gas krypton, and finally arriving at our current definition, which involves the distance light travels in a fantastically small, though extremely precisely defined, span of time. In this way, the speed of light and the length of the meter are now inextricably and definitionally linked. As our measurements become more precise, the speed of light doesn't change; instead, the definition of the meter does.

THE world of measurement involves much more than just the meter. If you wish to see how far down the rabbit hole of measurement it is possible to go, I recommend the *Encyclopaedia of Scientific Units, Weights, and Measures: Their SI Equivalences and Origins.* Compiled by François Cardarelli, a French Canadian chemical engineer, it is truly a wide-ranging document. It has conversion tables for units of measurement throughout history, from Abyssinian units of length to Egyptian units of weight, from the Roman system of distance, which uses a *gradus* (a single stride when walking) and *passus* (two strides) to denote distance, to the Assyrio-Chaldean-Persian measurement system. This book is exhaustive.

Interested in moving beyond light-years and parsecs (about three and a quarter light-years) to describe outer space? Then consider the siriusweit, which is equal to five parsecs. Or wondering about the details of the fothers, a British mass for trading lead bullion, or the *kannor*, a Finnish unit of volume? This book can fulfill your needs.

There are even various discrete units included, such as the perfect ream, which is 516 sheets of paper, and the warp, which is four herrings; it is used by British fishermen and old men at kiddush.

In addition to all this useful and possibly not so useful information, the book includes an intriguing table that shows how each definition of the meter reduced errors in measurement overall: Each successive definition made the meter a bit less uncertain. On the facing page is a chart that displays the table in graphic form.

These data points aren't just for years. Each redefinition oc-

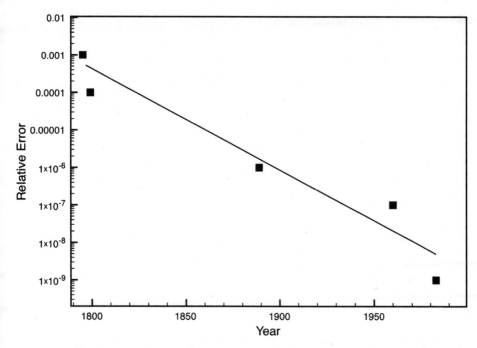

Figure 8. Measurement error of the meter over time. The precision in the definition of the meter has increased, with an exponential decay in error (the error axis is logarithmic) over time. Line shows general downward trend. Data from Cardarelli, *Encyclopaedia of Scientific Units, Weights, and Measures: Their SI Equivalences and Origins* (Springer, 2004).

curred at a very precise date. We know the very day when the meter became as precise as what we have now. Furthermore, the precision of the meter has increased in a regular fashion, and its error has declined in a linear fashion: an exponential decay. Just as scientific prefixes have changed in an exponential fashion, allowing for more precise terminology, so has the way we define measurement itself.

The meter's shift in precision, as well as definition, is not an aberration. Scientists have been driving toward ensuring that metric units in general be based on physical constants of the universe. In addition to the meter's tie to the light-year, time too is known precisely: A second is defined in terms of the vibration rate of a certain type of cesium atom.

The last basic unit in the metric system to undergo this transition to definition in terms of physical constants is the kilogram. For a long time the official kilogram was defined as the weight of a physical cylinder of platinum and iridium stored in a basement vault outside Paris. Over the past few years, metrologists—the scientists preoccupied with matters of measurement—have been bandying about alternative definitions, such as the mass of a sphere of silicon with an exact number of atoms or a precise amount of electromagnetic energy. In October 2011, they finally convened outside Paris at the Twenty-fourth General Conference on Weights and Measures and decided on a definition based on a physical constant as the official description of a kilogram.

Even though our units of measurement have become unbelievably exact, we don't normally really reach that level of precision. A certain amount of error and uncertainty are baked into our daily lives. Despite the increased precision with which all these units are now defined, we still deal with a certain amount of uncertainty when making measurements. Most people use a fairly basic ruler, despite the recent advances in precision. I can distinctly recall the yardstick my family owned when I was growing up—it was so worn at the ends that it was probably missing nearly an entire inch. Whatever I measured was objectively wrong, but it was close enough for everyday purposes. Similarly, when we exchange money from one currency to another, we don't mind that these conversions are necessarily approximate, inaccurate by several thousandths of a dollar or euro.

But understanding why we have measurement error, and properly understanding precision, can help us better understand how facts change and how measuring our world can lead to changes in knowledge.

IN 1980, A. J. Dessler and C. T. Russell published a tongue-in-cheek paper in *Eos*, a journal of the American Geophysical Union. In it, they examined the estimated mass of Pluto's size over time. We still don't know Pluto's mass, at least not exactly. Since it vents gases, astronomers often have trouble telling its size, sometimes viewing its self-generated haze as part of the surface.

Since Pluto's first sighting, when it was judged to be about the size of the Earth, estimates of its mass have dropped greatly over time. Dessler and Russell explained this by arguing something simple: Pluto itself is actually shrinking. By fitting the curve of Pluto's diminishing size to a bizarre mathematical function using the irrational number π, they argued that Pluto would vanish in 1984. But don't worry! According to their function, Pluto would reappear 272 years later (its mass would go from being mathematically imaginary to real again).

Of course, this is ridiculous. A far more reasonable explanation is that our tools improved over time via the Moore's-like laws that inexorably improve technology, allowing us to resolve Pluto better. While there's still a certain amount of uncertainty in Pluto's mass, we now have a much better handle on this fact: There is a clear relationship between what our facts are, increases in technology, and increases in measurement.

When it comes to error, measurement revolves around two terms: *precision* and *accuracy.* Any measurement method inherently has these two properties, and it's important to be aware of them when examining the true value of something. We can understand precision and accuracy through a rather whimsical scenario.

Imagine we are trying to determine the position of a point on a far-off surface by using a laser pointer. We have two different people tasked with trying to locate this point by aiming the laser pointer at it: a young boy who is lazy, and an older man who is very careful in his measurements. The young boy, endowed with the steady hand of youth, is physically capable of pointing the laser exactly at the point on the wall. But he doesn't want to for very long, because in our rather contrived example, he is inherently lazy, so he always chooses to rest his laser-pointer arm on a nearby surface. In this case, no matter how many times this boy points the laser at the point, it is always lower than the point by a little bit, because he chooses to rest his wrist on a lower surface.

On the other hand, the old man tries his best. He points the laser exactly at the point, but due to his age he has a slight tremor. So the laser is always hovering around the point, within a certain range, but it is rarely exactly where it should be.

In case this hasn't yet become clear, the old man embodies accuracy and the young boy embodies precision. Precision refers to how consistent one's measurements are from time to time. If the true length of something is twenty inches, precision refers to how dispersed one's measurements will be around the true value. If one measurement method always yields values of twenty-five inches, while another measurement method yields values within half an inch of twenty inches, but they are all variable, the former method is more precise, even if its results are wrong.

Accuracy refers to how similar one's measurements are to the real value. If your measurements are always five inches too high, even if your measurements are very consistent (and therefore are highly precise), you lack accuracy.

Of course, all methods are neither perfectly precise nor perfectly accurate; they are characterized by a mixture of imprecision and inaccuracy. But we can keep on trying to improve our measurement methods. When we do, changes in precision and accuracy affect the facts we know, and sometimes cause a more drastic overhaul in our facts.

EVERYONE recognizes the periodic table. The gridlike organization of all the known chemical elements contains a wealth of information. Each square itself holds a great many facts. For each element, we get its chemical symbol, its full name, and its atomic number and weight.

What exactly are these last two? Atomic number is simple: It represents the number of protons in the nucleus of an atom of this element and, similarly, the number of electrons that surround the nucleus. Since the number of electrons dictates a large portion of the nature of the interactions of an atom, knowing the atomic number allows a chemist to get a reasonably good understanding of the chemical properties of the element quite rapidly.

The atomic weight, however, is a bit more tricky. When I was younger, in grade school, I learned that the atomic weight is the sum of the number of protons in the nucleus and the number of neutrons in a "normal" nucleus. However, we were also taught that the number of neutrons in each atom can vary. If an element, as defined by the number of protons, can have different numbers of neutrons in its nucleus, these different versions are known as *isotopes*.

Hydrogen, with its lone proton, is the normal isotope of hydrogen that we think of. However, there's another version, with one proton and one neutron, known as deuterium. If you make water using deuterium, it's known as heavy water, because the hydrogen is heavier than normal. Ice cubes of heavy water will actually sink in regular water.

So, really, what the atomic weight describes is something a good deal more complex than what I was told when I was young. The atomic weight is the "average" size of the nucleus, in proportion to the prevalence of all isotopes of that element in nature. So for element X, if there are only two isotopes, we take their relative frequency in the world and weigh these sizes accordingly. In doing so, the atomic weight yields a sort of expected weight of neutrons and protons if you were to take a chunk of that element out of the Earth and the isotopes are all neatly mixed in together.

For a long time, these atomic weights were taken as constant. They were first calculated more than one hundred years ago and propagated in periodic tables around the world, with the occasional updates to account for what was assumed to be more precise measurement. But it turns out that atomic weights vary. Which country a sample is taken from, or even what type of water the element is found in, can give a different isotope mixture.

Now that more precise measurements of the frequency of isotopes are possible, atomic weights are no longer viewed as constant. The Internal Union of Pure and Applied Chemistry recently acknowledged this state of the world, alongside our increased ability to note small variations, and scrapped overly precise atomic weights; it now gives ranges rather than specific numbers, although many periodic tables still lack them.

Through increases in measurement, what were once thought to be infinitely accurate constants are now far fuzzier facts, just like the height of Mount Everest. But measurement's role is not only in determining amounts or heights. Measurement (and its sibling, error) are important factors in the scientific process in general, whenever we are trying to test whether a hypothesis is true. Scientific knowledge is dependent on measurement.

IF you ever delve a bit below the surface when reading about a scientific result, you will often bump into the term *p-value*. P-values are an integral part of determining how new knowledge is created. More important, they give us a way of estimating the possibility of error.

Anytime a scientist tries to discover something new or validate an exciting and novel hypothesis, she tests it against something else. Specifically, our scientist tests it against a version of the world where the hypothesis would not be true. This state of the world, where our intriguing hypothesis is not true and all that we see is exactly just as boring as we pessimistically expect, is known as the *null hypothesis*. Whether the world conforms to our exciting hypothesis or not can be determined by p-values.

Let's use an example. Imagine we think that a certain form of a gene—let's call it L—is more often found in left-handed people than in right-handed people, and is therefore associated with left-handedness. To test this, we gather up one hundred people—fifty left-handers and fifty right-handers—and test them for L.

What do we find? We find that thirty of the fifty left-handers have the genetic marker, while only twenty-two right-handers have it. In the face of this, it seems that we found exactly what we expected: left-handers are more likely to have L than right-handers. But is that really so?

The science of statistics is designed to answer this question by asking it in a more precise fashion: What is the chance that there actually is an equal frequency of left-handers with L and right-handers with L, but we simply happened to get an uneven batch? We know that when flipping a coin ten times, we don't necessarily get exactly five heads and five tails. The same is true in the null hypothesis scenario for our L experiment.

Enter p-values. Using sophisticated statistical analyses, we can reduce this complicated question to a single number: the p-value. This provides us with the probability that our result, which appears to support our hypothesis, is simply due to chance.

For example, using certain assumptions, we can calculate what the p-value is for the above results: 0.16, or 16 percent. What this means is that there is about a one in six chance that this result is simply due to sampling variation (getting a few more L left-handers and a few less L right-handed carriers than we expected, if they are of equal frequency).

On the other hand, imagine if we had gathered a much larger group and still had the same fractions: Out of 500 left-handers, 300 carried L, while out of 500 right-handers, only 220 were carriers for L. If we ran the exact same test, we get a much lower p-value. Now it's less than 0.0001. This means that there is less than one hundredth of 1 percent chance that the differences are due to chance alone. The larger the sample we get, the better we can test our questions. The smaller the p-value, the more robust our findings.

But to publish a result in a scientific journal, you don't need a minuscule p-value. In general, you need a p-value less than 0.05 or, sometimes, 0.01. For 0.05, this means that there is a one in twenty probability that the result being reported is in fact not real!

Comic strip writer Randall Munroe illustrated some of the failings of this threshold for scientific publication: The comic shows some scientists testing whether jelly beans cause acne. After finding no link, someone recommends they test different colors individually. After going through numerous colors, from salmon to orange, none are found to be related to acne, except for one: The green jelly beans are found to be linked to acne, with a p-value less than 0.05. But how many colors were examined? Twenty. And yet, explaining that this might be due to chance does little to prevent the headline declaring jelly beans linked to acne. John Maynard Smith, a renowned evolutionary biologist, once pithily summarized this approach: "Statistics is the science that lets you do twenty experiments a year and publish one false result in *Nature*."

Our ability to measure the world and extrapolate facts from it is intimately tied to chances of error, and the scientific process is full of different ways that measurement errors can creep in. One of these, where p-values play a role, is when a fact "declines" (and sometimes even vanishes) as several different scientists try to examine the same question.

IN the late nineteenth and early twentieth centuries, astronomers obsessed over a question that was of great importance to the solar system: the existence of Planet X.

Within a few decades after the discovery of the planet Uranus in 1781—the first planet to be discovered in modern times—a great number of oddities were noticed in its orbit. Specifically, it deviated from its orbital path quite a good deal more than could be explained by chance. Scientists realized that something was affecting its orbit, and this led to the prediction and discovery of the planet Neptune in 1846.

This predictable discovery of a new planet was a great cause for celebration: The power of science and mathematics, and ultimately the human mind, was vindicated in a spectacular way. A component of the cosmos—complete with predicted location and magnitude—was inferred through sheer intellect.

Naturally, this process cried out for repeating. If Neptune too might have orbital irregularities, perhaps this would be indicative of a planet beyond the orbit of Neptune. And so, beginning in the mid-nineteenth century, careful measurements were made in order to predict the location and properties of what would eventually become known as Planet X.

But Planet X was a slippery thing. By the time Pluto was discovered (the presumed heir to Planet X, though Planet X is an anomaly distinct from Pluto), the mass of Planet X had been calculated no fewer than four times. As the estimating continued, based on aberrations in Neptune's orbit, it kept getting smaller.

The first estimate was in 1901, and it showed that Planet X should be nine times the size of Earth. By 1909, it was down to only five times the size of the Earth. By 1919, Planet X was expected to be only twice the Earth's mass.

Of course, we now know that Pluto, as mentioned earlier, is small compared to these estimates. While it's not likely to evaporate anytime soon, whatever Dessler and Russell might say, Pluto is far smaller than the expected Planet X.

So, even after Pluto's discovery, astronomers continued to examine the unexplained properties of Neptune's and Uranus's orbits, each time recognizing that Planet X didn't have to be as large as previously thought. Now the consensus seems to be that Planet X does not exist. Due to Voyager missions we now have a much better handle on Neptune's mass, and between that and other increasingly precise measurements, it seems that Planet X is not necessary to account for what was previously measured.

Unlike Pluto, which won't actually vanish, it seems that Planet X already has.

Such a story in the physical sciences—where certain effects and unexplained phenomena decrease in size over time—is rare. However, when it comes to the realm of biology, social sciences, or even medicine, where measurements are not always as clear, and the results are often much more noisy (due to messy issues such as human actions), this problem is much more common. It's known as the *decline effect*. In some situations, repeated examination of an effect or a phenomenon yields results that decrease in magnitude over time. In addition to facts themselves having a half-life, the decline effect states that facts can sometimes decay in their impact or their magnitude.

While some have made this out to be somewhat mysterious, that needn't always be the case, as shown in the example of Planet X. Increasingly precise measurement allows us to often be more accurate in what we are looking for. And these improvements frequently dial the effects downward.

But the decline effect is not only due to measurement. One other factor involves the dissemination of measurements, and it is known as *publication bias*. Publication bias is the idea that the collective scientific community and the community at large only know what has been published. If there is any sort of systematic bias in what is being published (and therefore publicly measured), then we might only be seeing some of the picture.

The clearest example of this is in the world of negative results. If you recall, John Maynard Smith noted that "statistics is the science that lets you do twenty experiments a year and publish one false result in *Nature*." However, if it were one experiment being replicated by twenty separate scientists, nineteen of these would be a bust, with nineteen careers unable to move forward. Annoying, certainly, and it might feel as if it were a waste of these scientists' time, but that's how science operates. Most ideas and experiments are unsuccessful. Scientists recognize that the ups and downs are an inherent part of the game. But crucially, unsuccessful results are rarely published.

However, for that one scientist who received an erroneous result (and its associated wonderfully low p-value), there is a great deal of excitement. Through no fault of his own—it's due to statistics—he has found an intriguing phenomenon, and quite happily publishes it.

However, should someone else care to try to replicate his work, she might find no effect at all, or if there is one, it is likely to be smaller, if the experiment should never have been a success in the first place.

Such is the way of science. A man named John Ioannidis is one of the people who has delved even deeper into the soft underbelly of science, in order to learn more about the relationship between measurement and error.

.   .   .

JOHN Ioannidis is a Greek physician and professor at the University of Ioannina School of Medicine, and he is obsessed with understanding the failings and more human properties of the scientific process. Rather than looking at anecdotal examples, such as the case of Pluto, he aggregates many cases together in order to paint a clearer picture of how we learn new things in science. He has studied the decline effect himself, finding its consistent presence within the medical literature. He has found that for highly cited clinical trials, initially significant and large effects are later found to have smaller effects or often no effect at all in a nontrivial number of instances.

Looking within the medical literature over a period of nearly fifteen years, Ioannidis examined the most highly cited studies. Of the forty-five papers he examined, seven of them (over 15 percent) initially had higher effects, and another seven were contradicted outright by later research. In addition, nearly a quarter were never even tested again, meaning that there could have been many more false results in the literature, but since no one's tested them, we don't know.

The research that had initially higher effects ranged across many areas of study. From treatment of HIV to angioplasty or strokes, none of these areas were immune to the decline effect. And, of course, a similar range of areas was affected by contradictions: coronary artery disease, vitamin E research, nitric oxide, and more. As the saying among doctors goes, "Hurry up and use a new drug while it still works."

What was the cause of the decline effect here? Did Ioannidis ascribe this to anything new? Far from being the result of anything spectacular or confusing, the decline effect often comes down to a matter of replication and importance. The more something is tested, the better we understand it. Often, more important areas are those that are tested more frequently. It is likely that there are a good deal more incorrect effects out there in the medical literature than we are even aware of, just waiting to be tested.

Of course, it's not always this clear. As Ioannidis noted:

> Whenever new research fails to replicate early claims for efficacy or suggests that efficacy is more limited than previously thought, it is not necessary that the original studies were totally wrong and the newer ones are correct simply because they are larger or better controlled. Alternative explanations for these discrepancies may include differences in disease spectrum, eligibility criteria, or the use of concomitant interventions.

We should be wary of jumping to conclusions.

Nevertheless, in consonance with the idea of increasing precision and p-values, Ioannidis wrote:

> In the case of initially stronger effects, the differences in the effect sizes could often be within the range of what would be expected based on chance variability. This reinforces the notion that results from clinical studies, especially early ones, should be interpreted using not only the point estimates but also the uncertainty surrounding them.

More recently, Ioannidis conducted the same test for various biomarkers and found that subsequent meta-analyses often found diminished effects. We must always be aware of the fact that we are dwelling in uncertainty. Forgetting can make us jump to unwarranted conclusions.

THESE contradicted effects are related to what is perhaps Ioannidis's most well-known paper, which has acted as a sort of broadside on many aspects of how science is done. His 2005 paper in the journal *PLoS Biology* was titled "Why Most Published Research Findings Are False." As of late 2011, it has been viewed more than four hundred thousand times and cited more than eight hundred times.

He lays out very clearly a mathematical argument for why many scientific claims are untrue. Elaborating on several of the themes already discussed, what he looks for are situations in which there are cases of false positives, instances where a finding is "discovered" even though it's not actually real.

In a wonderful bit from *The Daily Show*, correspondent John Oliver interviews Walter Wagner, a science teacher who tried to prevent, via lawsuit, the Large Hadron Collider from being turned on. The Large Hadron Collider is a massive particle accelerator capable of generating huge amounts of energy, and Wagner was concerned that it could create a black hole capable of destroying the earth.

When Oliver presses Wagner on the chances that the world will be destroyed, he states that "the best we can say right now is about a one in two chance." Wagner bases this on the idea that it will either happen or it won't, so therefore it must be 50-50.

But this is absurd. Prior to the testing of a hypothesis, there is a certain expectation of what might happen. As another scientist interviewed by *The Daily Show* stated, there is a 0 percent chance of the earth being destroyed, based on what we already know about the fundamental laws of physics and how particle accelerators work.

This probability—what we expect to occur when we test a hypothesis—is known as the *prior probability*. The prior probability is simply the probability that the hypothesis is true prior to testing. Once we've tested it, we then get something known as a *posterior probability*: the probability that it is true, after our test.

Ioannidis argues that in a given field there is a certain fraction of relationships between variables that are real but many more that are spurious. For each field, then, there is a ratio of the relationships that are real to those that aren't. Think of it as the ratio between smoking-causes-cancer hypotheses and green-jelly-beans-cause-acne hypotheses.

Ioannidis then uses this ratio, along with something known as our hypothetical experiment's *discriminating power*—a number that encapsulates the ability of the experiment to actually yield a positive result—to calculate whether the experimental result is valid.

Essentially, in a quantitative way, he shows that in a large number of situations—whether due to the study being done in a field in which the above ratio is fairly low, implying that the probability of a spurious relationship is high, or an experiment using very few subjects, or the study was done in an area where replication of results doesn't occur—statistically significant and publishable results can occur, even though they are actually not true.

Ioannidis helpfully provides a few corollaries of his analysis that grow out of common sense, and I've added my own annotations:

> *The smaller the studies conducted in a scientific field, the less likely the research findings are to be true.* If a study is small, it can yield a positive result more easily due to random chance. This is like the classic clinical trial joke, in which, upon testing a new pharmaceutical on a mouse population, it was reported that one-third responded positively to the treatment, one-third had no response, and the third mouse ran away.
>
> *The smaller the effect sizes in a scientific field, the less likely the research findings are to be true.* If an effect is small, it could be like Planet X, and we are simply measuring noise.
>
> *The greater the number and the lesser the selection of tested relationships in a scientific field, the less likely the research findings are to be true.* More experiments mean that some of them might simply be right due to chance, and get published.
>
> *The greater the flexibility in designs, definitions, outcomes, and analytical modes in a scientific field, the less likely the research findings are to be true.* If there's a

greater possibility of massaging the data to get a good result, then there's a greater chance that someone will do so.

The greater the financial and other interests and prejudices in a scientific field, the less likely the research findings are to be true. Since scientists are people too, and are not perfect beings, the greater the possible bias, the greater the chance the findings aren't true.

The hotter a scientific field (with more scientific teams involved), the less likely the research findings are to be true. More teams mean that any positive result gets a great deal of hype quite rapidly, and is pushed out the door quickly, but leads to research that can be easily refuted, with an equal amount of hype. Ioannidis refers to this as a cause of the Proteus phenomenon, which he defined as "rapidly alternating extreme research claims and extremely opposite refutations."

ONE simple way to minimize a lot of this trouble is through replication, measuring the same problem over and over. Too often it's much more glamorous to try to discover something new than to simply do someone else's experiment a second time. In addition, many scientists, even those who want to replicate findings, find it difficult to do so. Especially when they think a result is actually wrong, there is even more of a disincentive.

Why is this so? Regarding a kerfuffle about the possibility of bacteria that can incorporate arsenic into their DNA backbone—a paper published in *Science*—Carl Zimmer explains:

But none of those critics had actually tried to replicate the initial results. That would take months of research: getting the bacteria from the original team of scientists, rearing them, setting up the experiment, gathering results and interpreting them. Many scientists are leery of spending so much time on what they consider a foregone conclusion, and graduate students are reluctant, because they want their first experiments to make a big splash, not confirm what everyone already suspects.

"I've got my own science to do," John Helmann, a microbiologist at Cornell and a critic of the *Science* paper, told *Nature.*

Or to put it more starkly, as Stephen Cole, a sociologist of science at the State University of New York, Stony Brook, quoted one scientist, "If it confirmed the first researcher's findings, it would do nothing for *them* [the team performing the replication], but would win a Nobel Prize for *him*, while on the other hand, if it disconfirmed the results there would be nothing positive to show for their work."

But only through replication can science be the truly error-correcting enterprise that it is supposed to be. Replication allows for the overturning of results, as well as an approach toward truth, and is what science is ultimately about. In a paper that followed up on Ioannidis's somewhat pessimistic conclusion, researchers calculated that a small amount of replication can lead us to much more robust science. But how do we do this?

A number of scientists are trying to make it more acceptable, and easier, to publish negative results. Since science prioritizes the exciting and the surprising, it is nearly impossible to publish a paper that says that some hypothesis is false. In fact, unless the work overturns some well-known result or dogma, the publication will never receive a hearing. Many scientists are advocating for journals and databases devoted to publicizing negative results to fill this publishing void, and have begun such journals. These could act as a check on the positive results so often seen in the literature and help provide a handle on the nature of the decline effect. In addition, they have the potential to act as a series of guideposts for other scientists, allowing them to see what hasn't worked before so they can steer clear of unsuccessful research.

SCIENCE is not broken. Lest the above worry the reader, science is far from a giant erroneous mass. But how do we return from the brink, where error and sloppy results might appear to be widespread?

Luckily, many of the erroneous and sloppy aspects of science are rare. While they do occur in a few instances, science as a whole still moves forward.

As Lord Florey, a president of the Royal Society, stated:

> Science is rarely advanced by what is known in current jargon as a "breakthrough," rather does our increasing knowledge depend on the activity of thousands of our colleagues throughout the world who add small points to what will eventually become a splendid picture much in the same way the Pointillistes built up their extremely beautiful canvasses.

Science is not always cumulative, as the philosopher of science Thomas Kuhn has noted. There are setbacks, mistakes, and wrong turns. Nonetheless, we have to distinguish the *core* of science from the *frontier*, terms used by SUNY Stony Brook's Stephen Cole. The core is the relatively stable portion of what we know in a certain field, the facts we don't expect to change. While it's no doubt true that we will learn new things about how DNA works and how our genes are turned on and off, it's unlikely that the basic mechanism of encoding genes in DNA is some sort of mesofact. While this rule of how DNA contains the information for proteins—known as the central dogma of biology—has become more complex over time, its basic principles are part of the core of our knowledge. This is what is generally considered true by consensus within the field, and often makes its way into textbooks.

On the other hand, the frontier is where most of the upheaval of facts occur, from the daily churn in what the newspapers tell us is healthy or unhealthy, to the constant journal retractions, clarifications, and replications. That's where the scientists live, and in truth, that's where the most exciting stuff happens. The frontier is often where most scientists lack a clear idea of what will become settled truth.

As John Ziman, a theoretical physicist who thought deeply about the social aspects of science, noted:

> The scientific literature is strewn with half-finished work, more or less correct but not completed with such care and generality as to settle the matter once and for all. The tidy comprehensiveness of undergraduate Science, marshalled by the brisk pens of the latest complacent generation of textbook writers, gives way to a nondescript land, of bits and pieces and yawning gaps, vast fruitless edifices and tiny elegant masterpieces, through which the graduate student is expected to find his way with only a muddled review article as a guide.

And pity the general public trying to make sense of this.

The errors at the frontier are many, from those due to measurement or false positives, to everything else that this book has explored. But it's what makes science exciting. Science is already a terribly human endeavor, with all the negative aspects of humanity. But we can view all of this uncertainty in a positive light as well, because science is most thrilling and exciting when it's unsettled.

There is a sifting and filtering process that moves knowledge from the frontier to the relatively compact and tiny core of knowledge. We should enjoy this process, rather than despair. One of the most fulfilling aspects is not the upheaval and churning of facts, but rather being able to grapple with concepts that explain our world. And these new facts are now only possible due to measurement.

IN addition to exposing quantitative error or delimiting what's around us (as in the case of Mount Everest's height), measurement can also have the profound benefit of overturning simple ideas and creating new pieces of knowledge, things we never could have known before.

A stark example is that of war, and whether it exists. On its face, this is a silly question: Of course war exists. It has always existed and seems to be a regular state of affairs, at least somewhere in the world.

However, John Mueller, a professor of political science at Ohio State University, decided to actually test this notion. Mueller began with a careful definition of what a war consists of, which is a reasonably well-accepted definition that it is a conflict between two governments, or a government and an organized group (relevant for a civil war), in which at least one thousand people are killed each year as a direct consequence of the fighting.

Mueller compiled all of the data since 1946 from a variety of sources and showed that, after an increase from the beginning of the data set until the end of the Cold War, the number of wars has plummeted precipitously. He also showed that the vast majority of wars are civil wars. Aside from why this has occurred, the presence of war has now become a clear mesofact. In the past several decades, war has gone from being a common and growing occurrence to something that is quite rare.

This fact is astonishing. But it's also astonishing for a reason besides its counterintuitive aspect: This fact could not have been known without careful measurement. There is a burgeoning group of scientists who use such data, which has been derived through careful measurement, to discover whole new ways of thinking about our world. The patron saint of this kind of careful measurement is Francis Galton.

. . .

WHEN born in 1822, Francis Galton was the scion of an esteemed family. Galtons had been scientists and businessmen for generations, and his half cousin happened to be Charles Darwin. But Galton's own career did not get off to a particularly auspicious start. Despite being considered a prodigy at an early age, he had a rather peripatetic and undistinguished young adulthood, traveling the world and performing rather poorly at Cambridge University.

But beginning with the publication of the chronicles of these explorations, Galton burst forth as a scientist. Soon he was involved in fields ranging from biology and mathematics to photography and anthropology, making numerous contributions in each of these areas.

Running through all of his work was an obsession with data and numbers. Galton collected data on everything. He examined data on contemporary illustrious men. He looked at data on the performance of university students. He looked at people's heights. He wrote a paper entitled "On Head Growth in Students at Cambridge." He even wrote a letter to *Nature* about how people visualize numbers in their minds, based on his own research. Galton was the man who introduced fingerprinting to Scotland Yard. He even constructed a map of beauty in the British Isles, based on how many pretty women he encountered in various locations. Stephen Stigler, a statistician at the University of Chicago, has argued that Galton was the man who ushered in the Statistical Enlightenment.

Derek de Solla Price, a kindred spirit when it came to data, wrote the following of Galton:

> Galton's passion shows itself best, I feel, in two essays that may seem more frivolous to us than they did to him. In the first, he computed the additional years of life enjoyed by the Royal Family and the clergy because of the prayers offered up for them by the greater part of the population; the result

was a negative number. In the second, to relieve the tedium of sitting for a portrait painter, on two different occasions he computed the number of brush strokes and found about 20,000 to the portrait; just the same number, he calculated, as the hand movements that went into the knitting of a pair of socks.

Galton was not a man to shy away from data. While many of his results may no longer be accepted, he combined analysis and mathematical techniques to great effect, and in so doing, brought many new facts to light, facts that could only be learned through careful, exhaustive, tedious measurement.

Such a preoccupation with allowing data to reveal new facts is the hallmark of science. This obsession with great amounts of data is not an isolated incident, something specific to Galton and an aberration along the trajectory of science. Instead, it is part of a grand tradition that, especially in the fields of sociology and social psychology, has unleashed a great many intriguing and clever experiments.

Stanley Milgram, known for his shock experiment that explored obedience, and for being the first to measure the six degrees of separation, conceived of numerous elegant experiments. One of these has the whiff of Galton. Known as the sidewalk experiment, he had graduate students stand on a New York City sidewalk and look up. He then measured how many students were required for this group to get passersby to stop and join them, or at least to look up themselves. These data, carefully collected, brought certain ideas about collective behavior to the fore.

Others have done similarly odd social experiments in order to collect data. For example, researchers have examined the drinking establishment locations and characteristics in different communities, and even whether the elderly are capable of crossing the street in the time a given traffic light provides them.

In the past few years there has been a surge in what is being called *data science*. Of course, all science uses data, but data science is more of a return to the Galtonian approach, where through the analysis of massive amounts of data—how people date on the Internet, make phone calls, shop online, and much more—one can begin to visualize and make sense of the world, and in the process discover new facts about ourselves and our surroundings.

Simply put, measurement teaches us new things, things that we can only know because we have the tools to quantify our surroundings. However, measurement cannot be used in every case, with its light of quantification shining evenly on every surface. Some measurements are easier than others, and this unevenness can profoundly affect what we know.

WHAT can be measured, and when, affects what can be learned. If we can't measure something, this can actually create a bias in what we know. For example, in biology, there is something known as *taxonomic bias*. This is when we study certain livings things not because they are more prevalent but because we like them more, or because they are simply easier to find. Vertebrates—those animals that have a backbone and comprise most of the creatures we are familiar with—are the subject of the vast majority of scientific papers, despite being only a tiny fraction of the different types of animals on Earth. Sometimes, when this seems overly malicious—amphibians and reptiles getting less attention than birds and mammals, because they are slimier or otherwise less cuddly—some scientists even call it *taxonomic chauvinism*.

Far from being an obscure point in scientific knowledge, taxonomic bias can ripple outward into popular culture. Let's return to dinosaurs. These massive creatures, beloved by millions, are firmly ingrained in the popular consciousness. And none more so than the staples tyrannosaurus rex, triceratops, and stegosaurus: the usual suspects.

But there's a reason for this. When dinosaur paleontology began to truly take off in the United States, there were two main dig sites: Hell Creek in Montana and Como Bluff in Wyoming.

While Hell Creek was a somewhat later site, dug in the 1900s, Como Bluff in particular had an outsize influence, as it was the dig site of Othniel Marsh. And of course, Edward Cope contested this claim, making it another of the many battlegrounds in the Bone Wars.

These two sites have finds from the Jurassic and the Late Cretaceous. As some of the first big dig sites, due to being the easiest to work with and the most abundant in fossils, they hold a special place in American dinosaur history. What dinosaurs were first found, and found in abundance, at these sites? Tyrannosaurus rex, triceratops, stegosaurus. In fact, one of the main reasons that the brontosaurus rose to such prominence is due to Marsh's discovery of a complete "brontosaurus" skeleton at Como Bluff.

Today, any museum worth its voluntary admission price has a display featuring at least some of the big names of Dinosauria. Most people don't realize that it is for a simple reason: These were the easiest dinosaurs to find—the low-hanging terrible lizard fruit, as it were. They had the first-mover advantage in fossils, and have therefore gained an outsize share of our brains' stock of dinosaur knowledge. What we study is not always what is actually out there; it's often what we're interested in, or what's easiest to discover.

MEASUREMENT is a double-edged sword. It can create errors where none existed before. It can lead us to information about certain topics more frequently than we might have expected, creating a sort of informational bias. And it can create spurious scientific knowledge, facts that must be examined with some hesitation. While we might not be able to know at a glance which results will be wrong, there are scientific principles to understanding how measurement can lead us astray.

But measurement can also create new knowledge, whether overturning false facts or finding something out about our world we'd never known before. As our methods of measurement have improved—according to mathematical regularities—we are now in the position to know more about our world.

Measurement of our surroundings is an inherently human process. But separating facts from the people who make them, spread them, or debunk them is nearly impossible. And measurement is not even the half of it. Building on everything we have seen about how facts change, it's time to finally tackle that last realm: the human aspect of how knowledge changes.

CHAPTER 9
The Human Side of Facts

FROGS have a curious type of vision. If you hang a dead fly on a string in front of a hungry frog, it won't eat it. It is entirely unaware that it's there. But put a live fly into a room with a frog, and the frog will actively work on pursuing its lunch.

It seems that frogs can only see certain objects when they are in motion: If the food is not moving, it might as well not exist.

While this is not how human vision operates (in fact, I *don't* want my food to be moving), there is some truth in this for us. If something moves rapidly, or changes quickly, we notice it easily. But if there is only very slow change, we are often not aware of it. This is similar to the story I repeated of the frog that, if slowly boiled, will willingly submit to its own death. While there isn't any evidence of this, I, as well as many others, erroneously continued to mention this story. Why? Because I sensed some truth in it. From the weird "human statues" in parks whose slow movements we cannot see, to our inability to see plant growth, or our failure to wear layered clothing and be able adapt to slow temperature changes throughout the day, we are not well equipped to deal with slow change.

Humans are imperfect. We fall prey to optical illusions, heal better the more expensive placebo we receive, and often have faulty memories. Having evolved in East Africa, we are confronted on a daily basis with situations far removed from those we encountered as hunter-gatherers. We have a certain amount of evolutionary baggage that makes us ill equipped to deal with many aspects of modernity.

This does not mean that all of our choices are irrational or ridiculous, or that we should simply curl into a fetal position and give up dealing with our incredible world, which is so full of technology and modernity. But we do approach the world in what Dan Ariely calls "predictably irrational" ways. If we are aware of the quirks of our brains and psychology, however, we can better understand the decisions we make and the world we create for ourselves. Crucially for us, we approach new knowledge and the world of facts in a manner that is far from completely rational yet is still regular and predictable in its biases. This chapter is about how our brains, and our psychological quirks, affect the facts that each of us holds within our minds.

When John Maynard Keynes was asked about why he switched his position on monetary policy, he uttered the immortal, though likely apocryphal, bon mot: "When the facts change, I change my mind. What do you do, sir?" All too often, we don't act like Keynes. We get stuck in ruts and don't change, even when the facts change around us. Why is that? For a variety of reasons—whether we don't perceive the change, don't believe that change is occurring, or simply don't believe the facts—we are not perfect beings who adapt immediately to the changes around us.

This can be seen clearly when it comes to slow change over the course of our lifetimes, the mesofacts of life. Whatever the state of the world we are born into quickly becomes what we expect to be normal.

This condition is known as *shifting baseline syndrome*, and it refers to how we become used to whatever state of affairs is true when we are born, or when we first look at a situation. Since we are only capable of seeing change over a single generation, if slow change occurs over many lifetimes, we often fail to perceive it.

Shifting baseline syndrome was first identified and named by

Daniel Pauly to refer to what happened with fish populations throughout the world. When the Europeans first began fishing off Newfoundland and Cape Cod, fish were incredibly abundant. In the seventeenth century, the abundance of cod was "so thick by the shore that we hardly have been able to row a boat through them." It seemed as if nothing could ever deplete their numbers. But within less than two hundred years of fishing, many species were entirely wiped out.

How could this have happened? Pauly described the situation as follows:

> Each generation of fisheries scientists accepts as a baseline the stock size and species composition that occurred at the beginning of their careers, and uses this to evaluate changes. When the next generation starts its career, the stocks have further declined, but it is the stocks at that time that serve as a new baseline. The result obviously is a gradual shift of the baseline, a gradual accommodation of the creeping disappearance of resource species.

It's easy to remember what is normal and the "correct" state of affairs when we start something new, after having set our baseline, even if we only do this subconsciously. But we mustn't let that guide all of our thinking, because the result can be catastrophic.

Of course, shifting baseline syndrome can even affect us in smaller, more subtle ways. Alan Kay, a pioneering computer scientist, defined technology as "anything that was invented after you were born." For many of us, this definition of technology captures the whiz-bang innovations of the Web browser and the iPad: anything that appeared recently and is different from what we are used to. In this way, we fail to notice all the older but equally important technologies around us, which can include everything from the pencil to window glass.

But factual inertia in general, even within a single life span, is all around us. Ever speak with a longtime New Yorker and ask for subway directions? You'll be saddled with information about taking the IND, BMT, and IRT, when you were hoping for something that would mention a numbered or lettered train. These mysterious acronyms are the names of the agencies—Independent Subway, Brooklyn-Manhattan Transit, Interborough Rapid Transit—that formerly ran the subways in New York City. Despite the unification that began in the 1940s of these competing systems, many people still refer to them by their former names. Even if facts are changing at one rate, we might only be assimilating them at another.

Adhering to something we know (or at least knew), even in the face of change, is often the rule rather than the exception. On January 13, 1920, the *New York Times* ridiculed the ideas of Robert H. Goddard. Goddard, a physicist and pioneer in the field of rocketry, was at the time sponsored by the Smithsonian. Nonetheless, the Gray Lady argued in an editorial that thinking that any sort of rocket could ever work in the vacuum of space is essentially foolishness and a blatant disregard for a high school understanding of physics. The editors even went into reasonable detail in order to debunk Goddard.

Luckily, the *Times* was willing to print a correction. The only hitch: They printed it the day after Apollo 11's launch in 1969. Three days before humans first walked on the moon, they recanted their editorial with this bit of understatement:

> Further investigation and experimentation have confirmed the findings of Isaac Newton in the 17th century and it is now definitely established that a rocket can function in a vacuum as well as in an atmosphere. The Times regrets the error.

Why do we believe in wrong, outdated facts? There are lots of reasons. Kathryn Schulz, in her book *Being Wrong*, explores reason after reason why we make errors. Sometimes it has to do with our desire to believe a certain type of truth. Other times it has to do with being contrary (Schulz notes one surefire way of adhering to a certain viewpoint: Have a close relative take the opposite position). But oftentimes it is simply due to a certain amount of what

I dub factual inertia: the tendency to adhere to out-of-date information well after it has lost its truth.

Factual inertia takes many forms, and these are described by
the relatively recent field of evolutionary psychology. Evolutionary
psychology, far from sweeping our biases under the rug, embraces
them, and even tries to understand the evolutionary benefit that
might have accrued to what may be viewed as deficits.

So what forms can factual inertia take? Look to the lyrics of
Bradley Wray.

In December 2009, Bradley Wray was preparing his high school
students for a test in his Advanced Placement psychology class.
Wray developed a moderately catchy song for his students that
would help them review the material and posted a video of it online.

What was the topic of this song? *Cognitive bias.* There is a
whole set of psychological quirks we are saddled with as part of our
evolutionary baggage. While these quirks might have helped us on
the savannah to figure out how the seasons change and where food
might be year after year, they are not always the most useful in our
interconnected, highly complex, and fast-moving world. These
quirks are known as cognitive biases, and there are lots of them,
creating a publishing cottage industry devoted to chronicling them.

As sung by Wray, here are a couple (the lyrics are far from
Grammy quality):

> *I'm biased because I put you in a category in*
> *which you may or may not belong*
> *Representativeness Bias: don't stereotype this*
> *song. . . .*

> *I'm biased because I take credit for success, but*
> *no blame for failure.*
> *Self-Serving Bias: my success and your failure.*

These biases are found throughout our lives. Many people are
familiar with self-serving bias, even if they might not realize it: It

happens all the time in sports. In hockey or soccer, if the team wins, the goal scorer is lauded. But if the team loses? The goalie gets the short end of the deal. The other players are the beneficiaries of a certain amount of self-serving bias—praise for success, without the burden of failure—at least that's how the media portray it, even if they are not subject to this cognitive bias themselves. There are well over a hundred of these biases that have been cataloged.

IN the 1840s, Ignaz Semmelweis was a noted physician with a keen eye. While he was a young obstetrician working in the hospitals of Vienna, he noticed a curious difference between mothers who delivered in his division of the hospital and those who delivered at home, or using midwives in the other part of the hospital. Those whose babies were delivered by the physicians at the hospital had a much higher incidence of a disease known as childbed fever, which often causes a woman to die shortly after childbirth, than the women delivering with midwives. Specifically, Semmelweis realized that those parts of the hospital that did not have their obstetricians also perform autopsies had similarly low amounts of childbed fever as home deliveries.

Ignaz Semmelweis argued that the doctors—who weren't just performing autopsies in addition to deliveries but were actually going directly from the morgue to the delivery room—were somehow spreading something from the cadavers to the women giving birth, leading to their deaths.

Semmelweis made a simple suggestion: Doctors performing deliveries should wash their hands with a solution of chlorinated lime beforehand. And this worked. It lowered the cases of childbed fever to one tenth the original amount.

However, rather than being lauded for an idea that saved lives for essentially no cost, Semmelweis was ostracized. In the mid-nineteenth century, there was no germ theory. Instead, the dominant paradigm was a certain theory of biology that blamed disease upon imbalances of "humors." If you've ever noted that someone is in a "good humor," this is a vestige of this bygone medical idea. So the medical establishment for the most part ignored Semmelweis. This quite likely drove him mad, and he spent his final years in an asylum.

This tendency to ignore information simply because it does not fit within one's worldview is now known as the Semmelweis reflex, or the Semmelweis effect. It is related to its converse, confirmation bias, where you only learn information that adheres to your worldview.

The Semmelweis reflex and confirmation bias are important aspects of our factual inertia. Even if we are confronted with facts that should cause us to update our understanding of the way the world works, we often neglect to do so. We persist in only adding facts to our personal store of knowledge that jibe with what we already know, rather than assimilate new facts irrespective of how they fit into our worldview. This is akin to Daniel Kahneman's idea of *theory-induced blindness*: "an adherence to a belief about how the world works that prevents you from seeing how the world really works."

In general, these biases are useful. They let us quickly fill in gaps in what we don't know or help us to extrapolate from a bit of information so we can make quick decisions. When it comes to what we can literally see, our ancestors no doubt did this quite often. For example, we could have expected the top of a tree to look like other trees we have seen before, even if it were obscured from view. If it didn't look right, it should still fit into our mental worldview (for example, it looked strange because there was a monkey up there). But when it comes to properly evaluating truth and facts, we often bump up against this sort of bias.

The Semmelweis reflex is only one of many cognitive biases, and it is related to another problem of our mental machinery: *change blindness*. This refers to a quirk of our visual-processing system. When we concentrate on one thing or task very intently, we ignore everything else, even things that are important, or at the very least, surprising. A series of seminal experiments were done in this field by Christopher Chabris and Daniel Simons, professors at Union College and the University of Illinois, respectively. You've probably seen their experiments, in the form of fun little videos online.

In one, subjects are shown a video of individuals in a gymnasium. The people in the video begin passing basketballs to one another and the subjects are supposed to keep track of the types of passes (such as bounce passes) or who passes to whom, since the players have different colored jerseys.

Then something intriguing happens. Partway through the video, a woman dressed in a full-body gorilla suit walks among the basketball players. She stops in the center, beats her chest in true gorilla style, and continues walking through the players. Of course, this is surprising and strange and all kinds of adjectives that describe something very different from a normal group of people passing basketballs to one another.

But here's the startling thing: 50 percent of the observers of this video miss the gorilla entirely. This change blindness, also known as inattentional blindness, is a quirk of our information-processing system. When looking for one thing, we completely ignore everything else around us.

This bug is turned into a feature by magicians, who exploit our change blindness through the use of misdirection. A magician gets you to concentrate on his left hand, while the right hand is doing all the important sleight-of-hand. This kind of thing can even fool trained magicians, who are trying to learn the illusion.

One common way that magicians learn a new trick is through an instructional video. The magician will show you the trick, through the eyes of the spectator, then explain it, show it again, and then show it from a different perspective, or at least more slowly. An illusion that I once observed involved the use of a thumb tip—a false rubber thumb that can be used to conceal various objects, such as handkerchiefs. After the magician showed the trick, he informed the viewer that he made sure to make it easy to follow by using a bright red thumb tip.

Upon hearing this, I was shocked. I had been so focused on all the other aspects of the illusion, and the magician's use of misdirection, that I had entirely missed what was right before my eyes: a ridiculous red artificial thumb. I had been a victim of change blindness.

Change blindness in the world of facts and knowledge is also a problem. Sometimes we are exposed to new facts and simply filter them out, along the lines of the Semmelweis reflex. But more often we have to go out of our way in order to learn something new. Our blindness is not a failure to see the new fact; it's a failure to see that the facts in our minds have the potential to be out-of-date at all. It's a lot easier to keep on quoting a fact you learned a few years ago, after having read it in a magazine, than to decide it's time to take a closer look at the current ten largest cities in the United States, for example, and notice that they are far different from what we learned when we were younger.

But whichever bias we are subject to, factual inertia permeates our entire lives.

A clear example of how we often neglect to respond to change is when it comes to writing the date or the year on documents.

Have you ever written the wrong year during the first weeks of January? This happens in everything from homework assignments to legal documents. This can even happen in the extreme. On May 24, 2011, President Barack Obama visited the United Kingdom. While making a stop at Westminster Abbey, Obama decided to sign the guestbook. He wrote a very nice little note about the special relationship that the United States shares with Great Britain. The only problem was that he dated his signature May 24, 2008. Perhaps he hadn't had to write the date since he won election three years earlier. Either way, the inability to respond to change is not an issue only for the everyday; it reaches all the way to the top. Thankfully, even the law has taken into account our foibles and our inability to always update our facts. In courts, intent is what matters, and not unconscious muscle memory, so if you do this on a legal document, you're generally fine.

I decided to conduct a simple experiment to actually get a handle on people's factual inertia. To do this, I used a Web site created by Amazon called Mechanical Turk. The label *Mechanical Turk* derives from a well-known hoax from the eighteenth and nineteenth centuries. The Turk was a complex device that was displayed all throughout Europe. While appearing to be a chess-playing automaton, the Turk actually had a person in a hidden compartment, controlling the machine.

In homage to this, Amazon named its online labor market—a clearinghouse for simple tasks humans can easily perform but computers cannot—Mechanical Turk. These tasks include things like labeling photographs when they are posted, and Turkers, as the laborers are called, will often solve these problems for pennies. Mechanical Turk has recently become a wonderful test bed for social science experiments, due to the large supply of subjects, the low wages required, and the fast turnaround time for running an experiment. While certainly not a perfectly representative distribution of humanity, it is much better than most traditional experimental populations, which are generally college undergrads.

As part of my scientific research, I've been part of a team that has developed software infrastructure for running online experiments to see how people cooperate in networks on Mechanical Turk. But due to this, I have also gained an appreciation for Mechanical Turk. To get a sense of people's factual inertia, I thought it would be a good place to quickly survey a population about their beliefs and knowledge.

I decided to examine people's knowledge of best practices when it comes to treating a nosebleed. I can distinctly remember a nosebleed of mine about ten years ago, when my nose started bleeding spontaneously one evening. With one hand holding a tissue to stanch the blood flow, I used the other hand to search online for how to properly treat it. Do I lean back? Do I lean forward? Where do I pinch my nose? I had heard so much competing information that I really didn't remember the "right" thing to do any longer. Searching online mid-nosebleed was my only re-course.

So I put a similar question to my Turkers:

"If you have a nosebleed, what's the best way to handle it?"

A. Lean your head back
B. Lie down
C. Lean your head slightly forward
D. Lean your head all the way forward

"And how would you hold your nose?"

A. Hold nose completely closed
B. Pinch bridge of nose

I asked for one hundred responses and offered five cents for each survey completion. Even with paying the overhead to Amazon, I was paying less than six dollars for new knowledge. I had my first response in less than a minute, which incidentally was incorrect. After a couple of days, all the responses were in.

For the record, according to WebMD, the proper procedure is to lean your head forward and pinch the bridge of your nose, so the correct answers should be C (and perhaps D), and B.

But how did my subjects respond? Here's the breakdown for the Turkers:

"If you have a nosebleed, what's the best way to handle it?"

Lean your head back: 50
Lie down: 14
Lean your head slightly forward: 31
Lean your head all the way forward: 5

"And how would you hold your nose?"

Hold nose completely closed: 17
Pinch bridge of nose: 83

While the Turkers seem to have generally grasped the idea that holding your nose completely closed during a nosebleed is a bad idea, they do not know the proper position to assume, with only about a third correctly identifying the way to position oneself.

The results of this experiment are not terribly surprising. In addition to all of the cognitive biases that we are saddled with, it is difficult for us to keep abreast of all the information around us. When we are young, we are treated as little generalists, absorbing all manner of information. We learn geography, history, mathematics, how to read a map, and lots of science trivia. We are even able to learn entire languages relatively effortlessly.

But then, as we get older, a curious thing happens with our approach to education. In addition to no longer being compelled to learn all manner of things (because we are, after all, adults, and we really can't be compelled to learn anything at all), if we do continue to educate ourselves, we focus. We choose a major and learn all that there is to learn about a single topic, such as biology. Then we become experts in that area, well aware of all of the nuances, debates, and changes in facts within that field. We learn more and more about less and less.

But all of our earlier knowledge remains in stasis. Instead of it all growing and developing in a rigorous fashion, like whatever we choose to make our careers in, it generally stays the same. Unless we happen to stumble upon an article in a magazine or newspaper about a certain scientific finding, or unless something is so important and earth shattering that we can't help but remark upon this new fact's novelty, we remain stuck at the factual level of our grade-school selves.

We continue to refer to different countries as First World, Second World, or Third World, not recognizing that these terms refer to alignments in the Cold War. Or our awareness of the periodic table remains stuck at high school levels, not realizing that the number of elements has grown a lot since we were in chemistry class.

But this description of how we learn things when we're young, and then stop learning, is a little too oversimplified and straightforward. It turns out that many of us do update these sorts of facts, but it often happens only in bursts. And these jumps occur at precise intervals: the length of a single human generation.

THE science writer Brian Switek pointed out to me that when most people learn about a topic—we were chatting about dinosaurs, but it works for most anything—we learn it when young, in the time of our lives when obsessive knowledge-gathering is the default mode, and then we leave it aside as we turn to more mature topics, or simply other things that we are now interested in.

But we do return to the subject, if only when our own children have reached the same point. Rather than seeing these mesofacts change slowly, in a relatively smooth advancement of knowledge, you only encounter them in bursts, when the next generation does, such as when your child comes home and informs you that dinosaurs were warm-blooded and looked like birds. This *generational knowledge* appears staccato, even though the knowledge changes and accretes steadily.

Whatever doesn't conform to your childhood, and especially when it comes to dinosaurs, often seems wrong. Just as shifting baseline syndrome makes us assume that whatever state of affairs we were born into is the normal one, we don't often confront changing facts until another generation grows up with a different baseline. We are then forced to confront the difference between them.

This is true of what's currently happening with Pluto. If you ask young children in 2012 to name the planets, they go up to Neptune, and they finish by saying that Pluto is a dwarf planet, or distinguish Pluto in some other way. But this is likely a temporary condition. Those teaching want these kids to know about Pluto and its curious status. But soon enough, it will just fade away into a strange footnote, paralleling what happened back in the nineteenth century: Just as Ceres and the other large asteroids were once counted as planets (marked as such on charts and taught to schoolchildren for decades) until the discovery of the abundant minor planets of the asteroid belt, Pluto's special place will likely fade away.

Of course, what *generation* means needn't be literal, although it is often the case that the facts in our brain—and their lifetime—are tied to childbirth. We can also understand what a generation is more figuratively. For example, when it comes to university-specific knowledge, a generation time is far closer to four years than multiple decades, due to the turnover of students. Institutional memory, and its attendant facts and knowledge, are only as permanent as its generation time.

This was made clear to me when reading an essay by Michael Chabon. He was bemoaning the recent commercialization and corruption of the purity that is Lego. He began with noting how the Lego sets that children have nowadays are fraught with pieces of every color of the rainbow: pink, purple, sky blue, and more. Furthermore, there are themed sets, from Harry Potter and *Star Wars*, replete with specialized pieces. But back in his day—and I read this approvingly—there were only a small handful of colors: red, blue, green, black, white, and yellow.

Then he continued by lamenting the cause of the downfall from Lego's pristine nature: the minifigure. Those small people with the yellow faces and simple grins, who, as Chabon argued, have constituted nothing more than a bastardization of the Lego aesthetic.

Suddenly I was no longer in agreement with the essay. This was wrong. While I don't care for the themed and specialized sets, which even include a set that seems to revolve around alien abduction, I grew up with these minifigs. They were a part of my childhood! How dare Chabon view these elemental little men as a corruption of the Lego ideal? They are part of Lego's nature.

And this is precisely the problem. We both were simply rejecting anything newer than our own childhood. Just as many of us only view "technology" as anything invented after we were born, we took our baseline—when we started playing with Lego—as the way things should be in the realm of Lego.

But is it always this simple? Are we forever mentally stuck in whatever state of the world we were born into? Or can we change the knowledge in our heads, even if it's a bit harder?

IN chapter 5, I explored how knowledge spreads and diffuses. But even if it spreads rapidly, what about the speed with which it comes to be accepted? Just as there are phase transitions when it comes to what we know, there can also be phase transitions in how knowledge is accepted and assimilated. Because even when facts spread, sometimes they take time to actually fix in our minds. And this is just as true in the realm of the scientist as it is in the world of the layman.

Clearly, science is not an abstract venture that is done in isolation from everyday human issues. It is not some endeavor immune to passions and biases. Science is an entirely human process. Science is done through hunches and chance recognition of relationships, and is enriched by spirited discussion and debate around the lab. But science is also subject to our baser instincts. Data are hoarded, scientists refuse to collaborate, and grudges can play a role in peer review.

The human aspect of science plays an important role when it comes to the acceptance of new knowledge. We don't always weigh the evidence for and against a new discovery or theory and then make our decision, especially if it requires a wholesale overhaul of our scientific worldview. Too often we are dragged, spouting alternative theories and contradictory data, to the new theoretical viewpoint. This can be very good. Having more than a few contrarians keeps everyone honest. But it can also be very bad, as when Semmelweis was ignored and essentially driven mad by his colleagues' refusal to accept the truth. But eventually, in the face of overwhelming evidence, the majority will generally accept the new theory, before their recalcitrance becomes too counterproductive.

Lant Pritchett, a professor of international development at Harvard's Kennedy School of Government, is all too aware of this. In the field of international development there are many sacred cows, and challenges to them are not met with as much cool and calculating logic as one might wish. Pritchett recently proposed an intriguing idea to help developing countries: create lots of guest worker programs. But is everyone simply weighing its merits? Not exactly. Pritchett argues that a more apt way to describe how these ideas are adopted is that they often follow this trajectory: "Crazy. Crazy. Crazy. Obvious."

Plot that on a graph, and you've got a phase transition, but this time it's one about how ideas are accepted and adopted. Thomas Kuhn, a physicist turned historian of science, also discussed how such rapid transitions occur in his celebrated book, *The Structure of Scientific Revolutions*. Kuhn used the term *paradigm* to refer to a holistic worldview or theory that can be used to explain our surroundings. (While Kuhn did not invent the word *paradigm*, he used it so much and so often that he is credited with its popularization.) For example, Newtonian gravitation is a very good theory, and has a great deal of explanatory power. But while Newtonian mechanics is actually still used for a large number of engineering applications, it has since given way to the theoretical worldview put forth by Albert Einstein. This change in perspective was termed a *paradigm shift* by Kuhn.

Kuhn argued that switching from one paradigm to another is a messy process and often involves scientists digging in their heels to the extent that their retirement or death—with their attendant replacement by younger and more open minds—might be required for the new paradigm to become accepted.

Maxwell Planck, another physicist, codified this in a maxim: "New scientific truth does not triumph by convincing its opponents and making them see the light, but rather because its opponents eventually die, and a new generation grows up that is familiar with it."

This seems intuitively obvious. Due to science being the biased and human affair it is, we can't expect the old stalwarts of science to change their minds when a new idea comes along. We just have to wait for them to die.

However, Planck's Principle turns out to be wrong.

This can be seen through a careful examination of the work of Charles Darwin. The quintessential phase transition in science, and paradigm shift, is that of the theory of evolution by natural selection. Everything in biology prior to evolution was sophisticated stamp collecting, ordering the living world around us and exploring its wonders. With the advent of evolution, biologists finally had a conceptual framework to make sense of the facts surrounding them. But the acceptance of evolution wasn't immediate. While *On the Origin of Species* was a bestselling book, it did not find universal agreement within the Victorian populace.

The same was true of the scientists themselves. David Hull, a philosopher of science, examined many of Darwin's well-known contemporaries to see who eventually accepted the theory of natural selection, and how long it took them to do so. Hull, along with two graduate students, Peter Tessner and Arthur Diamond, examined sixty-seven British scientists from Darwin's time. They found that only about three quarters of them had accepted Darwinian evolution ten years after *On the Origin of Species* was first published in 1859. So evolution was *not* the rapid phase transition of knowledge acceptance we thought it might have been.

But is this due to the vast majority of the holdout scientists being older? Were Darwin's ideas rapidly accepted by the younger generation, and was age simply masking what was in fact a phase transition among the younger scientists? It is true that the average age of those who accepted evolution was younger than those who still rejected it after ten years. But there are some complications. Age explains only about 5 percent of the variation of acceptance or rejection of this theory. The younger scientists didn't necessarily accept it rapidly; they accepted it at a rate similar to the older scientists who accepted it, over the course of a decade. More recent research into Planck's Principle has generally confirmed Hull's initial insight: Planck's Principle doesn't hold. Younger scientists aren't necessarily more likely to accept new ideas, and new ideas don't spread through a population as rapidly as we might expect.

While there are biases for how we assimilate facts, we can't even rely on common sense for understanding how factual inertia works: We have to test our irrationality. This is encapsulated in the work of Duncan Watts, a principal researcher at Microsoft Research. Watts has demonstrated, in numerous studies that explore everything from how certain songs become popular to how marketing works, that we are very good at telling stories to ourselves that sound true but must be subjected to the rigors of quantitative analysis for verification.

Understanding how concepts penetrate a group's consciousness in the scientific realm combines both the spread of knowledge through a population and all the cognitive biases we've discussed so far. But only looking at a single discipline, like biology or even economics, doesn't quite capture how prevalent the issues are that affect the delicate interplay between individual beliefs or ideas and the overall "facts" of a community.

One area in which we are forced to grapple with all of this interplay in its wonderful complexity—between what the community knows and what each of us knows—is in the realm of language.

LANGUAGE is a fickle thing, always changing. This is even recognized in the two ways linguists discuss grammar: *prescriptive grammar* and *descriptive grammar*. Prescriptive grammar is the way things ought to be, while descriptive grammar is the way things are. Prescriptivists held sway in centuries past, declaring what is allowed and what is not. They are responsible for such blanket rules as bans on split infinitives or ending sentences with prepositions.

On the other hand, descriptivists aim to chronicle the way we actually use language. While it turns out that we are still subject to many rules, these are often subconscious and less set in stone.

There can be a great deal of overlap between these two areas of grammar, but it diminishes as time goes on, as our actual language shifts and changes around us, widening the gap between the stone-inscribed rules of the prescriptivists and the observations of the descriptivists.

Language is a complex mix of flux and stability. On the one hand, there is evidence that the frequencies of the sounds of consonants in Old English are by and large the same as those in modern English, even though we modern English speakers are separated from Old English by one thousand years.

On the other hand, we also have many cases of linguistic change, such as new words being introduced and old words going extinct. Similarly, words themselves change, such as when verbs become more regular over time, and become more adherent to grammatical rules. In English we have verbs that are both regular and irregular. For example, the past tense of *discuss* is *discussed* (a regular verb that fits the "-ed" past tense), but the past tense of *speak* isn't *speaked*; it's *spoke*. Luckily, this change is not random: It turns out that the more frequently used words are those that are less likely to change, with a clear quantitative rule. Specifically, the rate of a verb's regularization is inversely proportional to the square root of its usage frequency. So how can we understand linguistic facts, and their interplay between change and stability?

Most of the facts we have examined so far are either what we as a society think is true (as in scientific truth) or what is the current state of the world (such as the speeds of the fastest computers). But when it comes to language, we're in a different sort of factual realm. Unlike those people who adhere to a prescriptive approach to linguistics, there's no real objective truth, with immutable rules that reside in some manual and that is completely independent of language speakers. A misuse of a word isn't wrong if enough people begin using it that way. Once most people start using *disinterested* and *uninterested* interchangeably, it just becomes annoying to continue to correct everyone.

The facts of language are a sort of population average of each individual's set of rules. Each person's approach, known by the delightful term *idiolect*, is a mercurial thing that is subject to what you learned when you were young, and to who's around you. It includes your vocabulary, grammar, pronunciations of words, and accent. Our linguistic facts hit the knowledge change jackpot: They are a complicated combination of slow adaptive change, factual inertia, and shifting baseline syndrome.

When speaking to others, we push and pull their speech patterns in various directions, even if only subconsciously, and they in turn influence us. There are many examples of this; one is *voice onset time*, which refers to how long it takes to produce the sound of certain consonants. This is performed completely automatically, but it is not unchangeable. While speaking with someone who has a longer voice onset time, a conversant often subconsciously begins to mimic the other.

Another subconscious language example is the *situation-based dialect*: A team of linguists studied Oprah Winfrey and how she introduced guests of different races. They found that she actually changed how she spoke during introductions, depending on whether her guest was white or black. This is similar to the person I know who was born in South Africa but raised in the United States: He only has a South African accent when speaking to his parents. Or how my wife switches between using *soda* and *pop*, depending on her location.

Of course, we're not entirely products of the influences of those around us; there are certain limits to our malleability. For example, while lengthening voice onset time causes a similar change in a listener, doing the opposite, shortening one's voice onset time, does not cause the conversant to also shorten theirs. Henry Kissinger, who has lived in the United States for well over seventy years, still has a very strong German accent; his idiolect has not changed one bit. Understanding language acquisition and change at the individual level is a complex and highly multidimensional issue.

But ultimately, seeing how language changes, and how we

mentally respond to this, can give us insights into how we adapt to the facts around us.

In an article about taboos and curse words, the linguist John McWhorter examines how this linguistic change happens around us:

> One reads with bemusement at scientists once perplexed at unearthing enormous bones of creatures now nonexistent. Between the teachings of the Bible and the brevity of a human life span, it took centuries to grasp that the world's fauna and flora have been in an eternal and imponderably long state of transformation. On language, the layman is today often in a similar state of perplexity. A language, too, is as inherently changeable as the lump in a lava lamp. However, print lends a sense that "real" language doesn't change, and we live too briefly to see much but hints otherwise.
>
> Hints, of course, we do see: When Ginger Rogers says in an old movie that a man "made love to" her we know she means what we would express as "come on to." However, we do not live long enough to know that two hundred years ago *obnoxious* meant "subject to injury" or that eight hundred years ago *quaint* meant "clever."

We are often like objects being dragged through mud. We change, but slowly, and with the residue of where we came from upon us.

Sometimes these changes are rapid and widespread, such as during the wonderfully named period in human history known as the Great Vowel Shift. The first time I stumbled across this phrase in my introductory linguistics textbook, I was fascinated. Apparently there were linguistic equivalents to the Black Death, the Great Awakening, the Enlightenment, and the Industrial Revolution. When I looked more carefully, though, it wasn't quite as dramatic as I first expected. While its exact causes are still unknown, it involved a shift, over the span of a couple of hundred years, beginning in the fourteenth century, when the pronunciation of certain vowel sounds changed. It is the reason that we now say "mouse" and "mice" instead of "moose" and "meese," which is what they used to be.

But imagine living during this. As people changed, what would be our response? Would we be confused by these changing facts or adapt rapidly to what was happening?

I had my own personal, far less great, example of this. It occurred when I was younger, when my brother and I were speaking with my grandfather. One of us described some activity as "very fun" only to have our grandfather inform us that this was not proper speech. One simply does not say that something is "very fun." But we felt that there wasn't anything wrong with it. I can

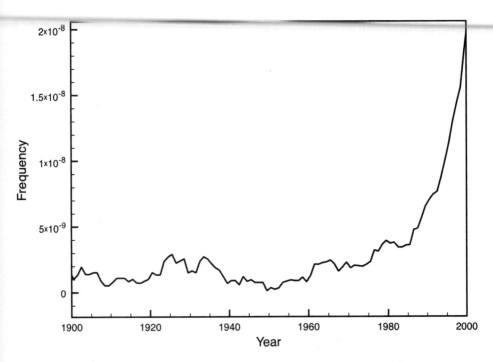

Figure 9. The frequency of the phrase "very fun" over time, as a curve. Notice that around 1980, the phrase's frequency increases rapidly. Data courtesy of Google Books Ngrams and the Cultural Observatory.

clearly remember our confusion, trying to tell our grandfather that people say this all the time, and that what we had said was correct.

In fact, we were simply part of a shift in usage that was happening around us. While considered improper English for nearly two hundred years, this phrase became acceptable around the early 1980s.

This is an example of shifting baseline syndrome and can perhaps give us a hint of what it would have been like to live during the Great Vowel Shift. My grandfather had not recognized the slow shift in language around him until confronted with generational knowledge, or, in this case, a two-generation jump in the linguistic facts around him.

This sort of shift in one's own mental linguistic rule set has actually been quantified in an attempt to understand it. When, and in what situation, we learn a language often affects how we process it for our entire lives, and sometimes even affects how we view the facts around us, such as in the case of my grandfather.

Linguists have looked at various aspects of a regional accent based on age. There are numerous examples of differences in the numbers of changes being present as a population is examined by age: The older the people examined, the less likely they are to have a certain linguistic innovation, whatever it may be. Of course, people also change during their own lives, too; despite numerous instances of people adhering to what they learned during childhood, there are also many instances when they alter their speech patterns both consciously and unconsciously over the course of their lives. The comedian and actor Stephen Colbert, for example, made a concerted effort to lose the Southern accent of his South Carolina childhood. But there are many times when linguistic change at the level of the community can be seen by looking at speakers of different ages.

Such linguistic decay is quite widespread. For example, something like the Great Vowel Shift occurred among French Canadians during the twentieth century. In Quebec, over the course of several decades, they began to change their pronunciations of the vowels in certain words. And just as before, those who were younger were more likely to exhibit this shift.

But, intriguingly, it didn't happen evenly across all words; there was a certain situational aspect to the shift. Words associated with the good old days, such as those having to do with parents, World War I, and even iceboxes, did not change alongside other words. There is speculation that younger speakers heard these words more often from their elders (who did not change their vowels), and were thus more likely to maintain the original pronunciations. One's linguistic facts are affected in a very real way by who we hear them from.

This is similar to my relationship with certain terms from cellular biology. When I took a cell biology course in college, I learned the topics from a British professor. Since I haven't gone on to study cell biology further, I am certain that many of the terms I use for cellular organelles are frozen in a British-style pronunciation. This was made clear when I was speaking with my father about programmed cell death, a process known as *apoptosis*. I said it as "a-puh-TOE-sis," only to be informed that many Americans actually say "a-pop-TOE-sis." While the American pronunciation sounded far sillier to me, apparently I was the one who sounded silly.

Ultimately, the facts of language—in this case, the prescriptive rules that we learn as schoolchildren—are those that are based on our own language experiences. Just as we view technological innovation, facts about dinosaurs, and even the acceptability of different types of Lego pieces with a perspective formed in the crucible of childhood, the same thing happens with language.

Whatever sort of knowledge we have, at this point it should be abundantly clear that we are far from perfect when it comes to having our own up-to-date facts in our minds. But must it be so? Or can we adapt to all of this change? Can we recognize all the biases that are part of each of us and prevent our knowledge from being updated only once a generation or so?

. . .

THERE are already institutions that deal with the far reaches of the fact timescale, trying to help us deal with change. On the fast end of facts, there is a company started a few years ago called Ambient Devices. A spin-off from the MIT Media Lab, Ambient Devices has created a number of informational appliances for rapidly changing information, including market and weather data. For example, they offer an orb that can be placed on a desktop that glows red when the market is down and green when the market is up. They offer the Ambient Umbrella, whose handle glows when rain is in the forecast, so it won't be forgotten. These gadgets provide a way for us to be kept abreast, in a vaguely useful way, of the changing facts around us. But these sorts of devices are only good for quickly changing facts.

On the other end of the spectrum there is the Long Now Foundation, which is geared toward fostering long-term thinking and awareness. They get people to think in terms of millennia, and are even constructing a clock out in the Texas desert. Designed to operate for ten thousand years. They also have a sporadically used news-clipping service that highlights articles that might be relevant centuries from now, as a way to keep on top of really big and important changes.

But what of everything else? What of all the facts that change at that intermediate rate, along the timeframe of years, or of a single generation? This is the timescale of the mesofact, and the one on which most of us fail to keep up-to-date with our facts, the one where cognitive failures reign.

There is a straightforward, though not always easy, solution to dealing with changing facts: constant education and the omnipresence of information. How each of us implements these rather difficult solutions is certainly a personal choice, but the following are a number of suggestions.

For example, there is a Web site called Worldometers. It is part of the Real Time Statistics Project and acts as a clearinghouse for the sorts of estimated real-time counters we have likely seen, such as the National Debt Clock near Times Square in New York City, which counts the U.S. debt. Worldometers aggregates counters from numerous organizations for such quantities as the current world population and the number of member countries in the United Nations, new books published this year, HIV/AIDS–infected individuals, the amount of coal left on Earth, and the number of species that have gone extinct this year. While many are clearly estimates, they are carefully curated and provide a measure of our current factual awareness of the world.

But what else can we do? It's a lot easier to choose a few things to focus on rather than drink from the unfiltered firehose of ever-changing facts. How do we balance this?

Perhaps some form of informational triage is in order. Or failing that, we can periodically require a radical reexamination of what we believe. How should we do this? Let's take the example of prices, facts that are always around us.

Most of us have an intuitive sense of the cost of everyday goods. When we hear that corn costs a dollar for twelve ears, we think this sounds low, and we will likely buy it. On the other hand, some of us might not be aware of certain prices. I'm lactose intolerant, so I have very little sense of the price of a gallon of milk. Frequently checking the Consumer Price Index instead of looking at the prices of individual items allows us to gauge price changes in a more consistent fashion and to put current shifts into historical context. For example, most changes to the CPI have only occurred in recent years, while prices for consumer products were actually quite stable for centuries prior to the twentieth century.

Similarly, I don't know the current cost of gold at all, except that it's rather high. However, I remember following it carefully when I was in middle school, looking every day at the Money and Investing section of the *Wall Street Journal* and seeing if it was on its way up or down.

While this is one solution to making sure we aren't caught unawares by facts, looking up prices on a daily basis seems onerous for most of us. And going even further and putting all of it into a historical context, while interesting, is a lot to ask. It would be great if many people used the Web site MeasuringWorth.com, which has data on the purchasing power of the British pound going all the way back to 1245; somehow, however, I don't see this site becoming particularly popular anytime soon.

Of course, if someone is an expert on a topic, he is required to keep abreast of related information. Gold traders have to follow its price. Medical facts, as mentioned many times, change and are overturned very often. In order for physicians to deal with this continually vibrating knowledge, they are regularly required to take courses called continuing medical education. Every few years they're required to get up-to-date on what is considered state-of-the-art. Doctors used to listen to audiotapes and then take a fill-in-the-bubble test. Now they can listen to CDs or even stream the lessons online, and then take an online exam. As we continue to be bombarded by information in our everyday lives, we must take this lesson from experts and actively learn all the time; otherwise we are doomed to be saddled with outdated mesofacts.

It's much harder to do this when one's livelihood doesn't depend, for example, on medical practices or the price of corn or gold. It's hard enough to have the newest knowledge in one's own field, but dealing with knowledge that's outside of one's area of expertise is even harder. Unless we want to make it our jobs to figure out how to invest, just use index funds and don't bother focusing too carefully on individual stocks.

Perhaps the same advice can be used for knowledge. Unless it's one's job to keep abreast of a certain field of knowledge, simply use the informational equivalent of index funds. But what are informational index funds? They are publications and Web sites that aggregate changing knowledge all in a single place. These include magazines, blogs, and the "What's News" column in the *Wall Street Journal*, among other sources.

While informational index funds can help, reading omnivorously is still important, and we have already been given some help with this. *The Atlantic* has begun running a series called Media Diet, which asks influential thinkers what they read and how they get their facts and news. These influential people, from Gay Talese and the newspapers he carefully reads to David Brooks and the blogs he frequents, give their informational diets to help guide others.

But when it comes to being aware of facts, there's actually an even better solution: Stop memorizing things and just give up. That sounds terrible, but it's not. Our individual memories can be outsourced to the cloud. Specifically, rather than relying on memorizing often out-of-date facts, and still usually only half-remembering them, embrace the idea that we have the Internet at our disposal, with search engines at our fingertips that enable us to search for any fact we need anytime.

This is already happening. A recent paper in the journal *Science* finds that people are coming to rely more and more on search engines rather than their own memory. When the study was released, many people fretted about this and how it is hurting our brains and making us dumber. While this is certainly a common argument, I took away the opposite conclusion. Paradoxically, by *not* relying on our own memories, we become more likely to be up-to-date in our facts, because the newest knowledge is more likely to be online than in our own heads. Medicine has exploited this idea through a constantly updated online medical reference site called UpToDate; looking something up guarantees the most current information.

Of course, it's good to keep a certain state of affairs about the world in our minds, but the more we look things up, the more likely we are to not be caught unaware when we encounter a new and startling idea.

EVER-CHANGING facts are all around us. But there is often a disconnect between the state of knowledge around us and what we hold in our minds. This disconnect can sometimes be quite large, and it can be due to a wide variety of reasons, from factual inertia to generational knowledge. Happily there are ways of doing our best to avoid being so surprised when we encounter a new bit of information. Some of this involves simply recognizing the systematic ways in which we update our own personal store of knowledge, or by using technological tools to help us stay a bit more up-to-date. But we can also be aided by something more general, to which this book has hopefully acted as a guide: recognizing the regularities in how knowledge changes around us.

CHAPTER 10
At the Edge of What We Know

AN Italian Franciscan monk named Luca Pacioli wrote a book on mathematics in the late fifteenth century. This textbook, *Summa de arithmetica, geometria, proportioni et proportionalita*, was one of the earliest books published after the advent of the printing press. While focusing on algebra and other aspects of math, there also was a significant section on accounting.

Pacioli chronicled double-entry bookkeeping—entering each entry twice, both as a debit and a credit, in order to reduce errors—which was the first time this method had been codified in print. This error-checking methodology, which was being used by Italian merchants, finally could be spread far and wide.

Mary Poovey, an English professor at New York University and the author of the monograph *History of the Modern Fact*, has argued that the modern conception of the fact, with its notion of objective reality that often goes hand in hand with a certain quantifiable quality, was first seen during the Middle Ages. Specifically, Poovey identified the "invention" of the fact with the advent of double-entry bookkeeping. Only with the introduction of this methodology in the fifteenth century or so, Poovey argued, did humans become accustomed to thinking in terms of bits of information in quite this way.

Of course, this doesn't quite ring true. As humans we have

been chronicling pieces of knowledge for millennia, if sometimes with a bit less objective mathematical truth than that found in a ledger. We can see numerous examples as far back as the ancient Greeks. Anaximander, a philosopher from Miletus in Greece, who lived in the late sixth century BCE, detailed a number of facts about the origins of human beings: Initially fish grew out of hot water and earth, and then humans eventually grew inside them until they were capable of surviving on their own. This is one of the reasons that Anaximander discouraged the eating of fish.

Or we can look to Empedocles, another pre-Socratic philosopher, who wrote of the fact that while water and wine mix together, water and oil do not. Or Anaxagoras, who wrote much about nature and included a discussion of weather in one of his works, where he noted that thunder was a clash of clouds and that lightning is due to the friction of the clouds, like how sparks are formed. Shooting stars are sparks shaken from the air. While the ancient philosophers worked to order their world, they were more than a little hit-or-miss.

The world was ready for new types of knowledge by the late Middle Ages. As the Middle Ages and its accounting systems gave way to the Renaissance, which in turn laid the foundations for the Scientific Revolution, facts were given a new sort of prominence.

As the scientific method was being codified, and our surroundings were subjected to experimental rigor the likes of which the world had never seen, facts were generated and overturned at an ever-increasing pace. Finally, the testable scientific fact had arrived. This is the critical insight of the Scientific Revolution: Science requires an idea to be refutable. It is not good enough for a concept to seem compelling; it must have the potential for a new fact to come along and render it false. As we have seen over and over, this not only can happen, but often is the rule rather than the exception. Kathryn Schulz, in *Being Wrong*, notes:

> This is the pivotal insight of the Scientific Revolution: that
> the advancement of knowledge depends on current theories

collapsing in the face of new insights and discoveries. In this model of progress, errors do not lead us away from the truth. Instead, they edge us incrementally toward it.

Ever since the end of the Middle Ages and the Scientific Revolution—in addition to the Industrial Revolution, which saw ever-changing technological facts—we have lived in this new and exciting realm of knowledge. But what is next for facts, for their creation and displacement? Since the sixteenth century, we seem to have been surrounded by knowledge that is changing faster and faster. Will this continue to occur? Are the half-lives of facts themselves shortening, and must we learn to adapt more and more rapidly? Or are we living in some temporal sweet spot for facts, the turning point after which their changes in at least some domains might actually be slowing down?

THROUGHOUT this book I have explored various trends that seem to be speeding up, whether it's technological innovation, scientific understanding, or even how information wings its way around the globe. But will we ever reach some sort of plateau? Or are we forever doomed to contend with everything around us changing faster and faster?

As I mentioned in chapter 6, the innovative pace of cities is ever quickening. In order to continue growing, cities seem to actually require periodic drastic innovative reboots—from advanced sewage systems to building methods that allow for skyscrapers—that are happening more quickly. For the first time in human history these rapid changes are occurring multiple times in a single generation, and they don't seem to be slowing down.

This parallels work that has examined the rate at which great ideas come about and are integrated into society. In 1971, a team of researchers compiled a list of the sixty-two greatest advances in the social sciences since 1900, in an effort to understand how these concepts were generated and what their properties were. Most important, they examined the delay until the impacts of these major advances became manifest. They found that it takes about ten years for them to reach penetration, but critically, this time has been shrinking in recent years. No doubt, if a similar study were done again, we might find even shorter delays.

Recent climate change has even changed facts around the globe that we thought were far from mesofacts. There are certain island countries in the Pacific Ocean that will likely vanish in the next few decades due to rising sea levels, raising bizarre questions related to statehood: Do submerged islands, even though part of the ocean, remain nation-states? The Maldives and Kiribati are in this strange place concerning their statehood, and not because of conquest or political reasons. These countries' physical existences might be changing, a state of affairs we have never had to grapple with before.

It would seem that, in these cases, new knowledge is being created more rapidly, spreading faster, and causing changes in things that we had not even realized were in the realm of mesofacts.

On the other hand, there seem to be examples where the change in facts is slowing down. While Moore's Law has had an incredible run that has lasted decades, and it has incorporated successive generations of technology, there are many who feel that it only has a couple of decades or so more to go. At that point, in the near future, we will start to bump up against limits imposed by physics, such as the size of atoms, which will ultimately limit how many components we can cram onto a circuit.

The same sort of limits could be argued to hold with transportation speeds. We have had an astonishing sequence of technologies that have allowed us to go faster and faster, but an exponential pace doesn't seem sustainable forever. While going to the moon for our lunch break sounds wonderful, it just isn't likely.

This does not mean that technological change stops adhering to mathematical rules. But in the long term they might very well be adhering to a logistic curve, with an eventual slowing toward a limit. We are simply in the fast-changing portion in the middle right now, so it is hard to see the eventual slowdown. That being said, as humans we are very good at being pessimistic and underestimating our ability for continued innovation. Even though each individual technology might reach its limits, a new one comes along so often to innovate around these limits that the change around us might not be slowing down for a long time to come.

But what of scientific knowledge? While we are nowhere near the end of science—the sum of what we don't know is staggering—we might very well be in a logistic curve of ever-changing knowledge as well, rather than one of exponential growth. One of the reasons I believe this could be true is simple: demographics. It seems unlikely that the rapid population growth will continue growing faster and faster. Whenever a country has become industrialized, its development has gone hand in hand with a drop in birth rate. Therefore, as the world as a whole advances technologically, population will cease to grow at the frenetic pace of previous decades and centuries. Combined with energy constraints—we are nowhere near our limits, but our energy resources are certainly not unbounded—exponential knowledge growth cannot continue forever. On the other hand, as computational power advances, computer-aided scientific discovery could push this slowdown far off into the future.

Nevertheless, there are regularities to factual change and growth: Facts will continue to grow and be overturned, albeit at a slower place, and we certainly do not seem to be leaving the exponential regime anytime soon.

But even if everything continues to grow rapidly, there might be certain limits to how we perceive this change and adapt to it.

WHEN Carl Linnaeus worked out his methodology for organizing all living things in the eighteenth century, his taxonomy had three kingdoms—roughly translated as animal, vegetable, and mineral—and further divisions into classes, orders, genera, and species. Biologists now have five kingdoms, new subdivisions between kingdoms and classes called *phyla* (singular *phylum*), families between orders and genera, and even three larger overarching divisions above kingdoms known as domains. As our knowledge has grown from thousands of species to millions, so too has our system of classification.

Similarly, the way we categorize different diseases has grown rapidly. In 1893, the International List of Causes of Death was first adopted and contained about 150 different categories. As of 2012, we are up to the tenth revision of the International Statistical Classification of Diseases and Related Health Problems, known as ICD-10. It was released in 1990 and has 12,420 codes, which is nearly double that of the previous revision, ICD-9, which came out only a little more than ten years before ICD-10. As facts have proliferated, how we manage knowledge and think about it has also had to grow, with our classification systems ramifying in sophisticated and complex ways.

On the one hand, being exposed to more complexity, whether it be in the realm of categorization of diseases, living things, or the many other classification systems we use—from types of occupations to Internet domain names—could make us more intelligent. Just as being exposed to cognitively demanding television shows and video games seems to increase our ability to think critically, so too could more facts, and their attendant complex classification systems, make us smarter.

However, as humans, we seem to have certain cognitive limits on what we can know and what we can handle in our daily lives.

Our brains are only so big. And it seems that the sizes of our brains actually dictate how many social connections we can have and how many people we can regularly interact with and keep in our minds. Dubbed Dunbar's Number, after its discoverer, Robin Dunbar, who examined the brain sizes of different primates, the number of people we can know and have meaningful social ties with seems to be limited to between about 150 and 200. This is about the number of soldiers that comprise a fighting unit—whether in ancient Rome or modern-day armies—and fits the size of a small village. Surprisingly, despite technological advancements in the social networking sphere, our number of Facebook friends still adheres to Dunbar's Number and is about 190, as of 2011.

Similarly, if we look at the number of close ties we each have, we discover another trade-off. While we know a lot of people, each of us really only has a handful of very close social ties, such as our spouse or best friend. For most people, this number is around four. In my own research I has found that as we increase the number of people we are close to, we lower how close we are to each of them, on average. So if I have five friends instead of four, I am less close to each of these five people than I would be if I eliminated one of them from my tight inner circle. There seems to be some sort of conservation of attention: As we increase who we pay attention to, we spread this amount of attention out evenly among these individuals.

Our brains have a certain capacity, at least when it comes to social ties. Is the same thing true for changing knowledge? Upon being confronted with his ignorance of the Copernican notion that the Earth orbits the Sun, Sherlock Holmes argued this very point:

"You see," he explained, "I consider that a man's brain originally is like a little empty attic, and you have to stock it with such furniture as you choose. A fool takes in all the lumber of every sort that he comes across, so that the knowledge which might be useful to him gets crowded out, or at best is jumbled up with a lot of other things, so that he has a difficulty in laying his hands upon it. Now the skilful workman is very careful indeed as to what he takes into his brain-attic. He will have nothing but the tools which may help him in doing his work, but of these he has a large assortment, and all in the most perfect order. It is a mistake to think that that little room has elastic walls and can distend to any extent. Depend upon it there comes a time when for every addition of knowledge you forget something that you knew before. It is of the highest importance, therefore, not to have useless facts elbowing out the useful ones."

We very likely can't handle every piece of knowledge that comes our way, and while being exposed to more and more might help us to think better, we no doubt have our limits when it comes to dealing with rapidly changing facts. This sounds like bad news. Our brains simply won't be able to handle all of this knowledge and information, and the rapidity at which it changes. There are workarounds, such as those mentioned in the last chapter, i.e., online search engines. But, happily, it turns out that even when rapid change happens, it's not as overwhelming as we might think.

Many futurists are concerned with what are termed *singularities*, periods of such rapid and profound change due to technology that the state of the world is forever altered. Like some of the changes mentioned in chapter 7, these phase transitions happen so quickly that they can forever alter humanity's relationship with its surroundings. The quintessential singularity that futurists dwell on is that of the potential creation of superhuman machine intelligence. While many scientists think this is either very far off or that it will never happen, how would singularities affect us? Would a singularity tax our cognitive limits or will we be able to cope?

Chris Magee, the MIT professor who studies the rapid technological change around us, and Tessaleno Devezas of the University of Beira Interior in Portugal, decided to use history as a guide. Focusing on two events that have already happened, Magee and Devezas decided to see how humanity has dealt with fast change. They first looked at how the Portuguese gained control over increasingly large portions of the Earth's surface over the course of the fifteenth century, as their empire grew. They also looked at the progression of humanity's increasingly accurate measurement of time over the last millennium or so. In both cases there were rapid shifts in certain facts, all according to exponentially fast growth and culminating in what many would argue was the crossing of some sort of singularity threshold. In the case of Portugal, the country established a nearly globe-encompassing maritime empire, and in the case of clocks, timepieces became so advanced that measurement of time was far more precise than human perceptions.

But humanity assimilated these changes quite well. When speaking about the innovation in timekeeping, Magee and Devezas wrote:

> These large changes were absorbed over time apparently without major disruption; for example, no mention is made of "clock riots" even though there was resistance and adaptation was needed. In given communities, the large changes apparently happened within less than a generation.

So I think it is safe to assume a somewhat optimistic tone, recognizing that change, while it might be surprising to many of us, is not entirely destabilizing. Humans are very adaptable, and are capable of understanding how knowledge changes.

And, of course, that's the message of this book itself.

As I hope I've shown, facts can change in a startlingly complex variety of ways. But far from the fluctuation in our knowledge being random, the changes are systematic and predictable. Whether about nature or about the man-made world, factual change due to measurement changes or even the identification of errors, facts change in recognizably regular ways.

In addition to looking up facts on the Internet, or to having glowing orbs on our desk that respond to changes in the market, another way to avoid the surprise of changes in knowledge is to simply recognize that it's not that surprising.

We are getting better at internalizing this. For example, many medical schools inform their students that within several years half of what they've been taught will be wrong, and the teachers just don't know which half. But too often—whether because change is still too slow to notice or because of quirks in how we learn and observe our surroundings—we don't really live our lives with the concept that facts are always changing.

In an interview, the novelist Jonathan Franzen noted: "Seriously, the world is changing so quickly that if you had any more than 80 years of change, I don't see how you could stand it psychologically." Many of us still maintain this attitude, unable to deal with change. But it doesn't have to be this way. We have to begin actually educating ourselves and our children to recognize that knowledge will always be changing and showing the regularities behind how these changes can happen. More important than simply learning facts is learning how to adapt to changing facts. Until we begin to do that, we are going to continue to be caught flat-footed by new information.

Facts don't change arbitrarily. Even though knowledge changes, the astounding thing is that it changes in a regular manner; facts have a half-life and obey mathematical rules. Once we recognize this, we'll be ready to live in the rapidly changing world around us.

ACKNOWLEDGMENTS

An unbelievable number of people have been instrumental in making this book a reality. In the world of science I have had a great number of supporters and mentors. While it would be nearly impossible to list everyone, I would like to single out Steve Strogatz and Nicholas Christakis. Steve, my graduate school adviser, is a great mentor and friend, and a collaborator on numerous research projects. In addition, he encouraged me in writing and even provided me with my first opportunity to write for a large audience, at the *New York Times*. Nicholas Christakis, whose group I was a part of while a postdoctoral fellow, has also been a wonderful collaborator and friend, as well as acting as a mentor in both my research and writing. I have been privileged to work with such amazing scientists and writers, who nurtured the highly interdisciplinary and unorthodox path I have chosen for myself.

I have also had a number of supporters in the writing world. Gareth Cook, my former editor at the Ideas section of the *Boston Globe*, discovered my early writing and nurtured my skills. Both Gareth and Steve Heuser, the current editor of Ideas, also presided over the publication of my article about mesofacts that first got this whole book-writing process going. Thanks go to both the *Boston Globe* and *The Atlantic*, where I was given the opportunity to write several articles that have been adapted here. David Moldawer, who bought this book when it was a mildly coherent shell of what it hopefully has become, also deserves my thanks.

I have had the pleasure of having many supportive readers of early drafts, who also brought various concepts and articles to my attention. Thank you to Avi Gerstenblith, Paul Kedrosky, Jukka-Pekka Onnela, Jason Priem, Niels Rosenquist, and Josh Sunshine.

K. Brad Wray deserves thanks for introducing me to so many ideas in the philosophy of science, including the fallacy of Planck's Principle, in addition to taking the time to read an entire draft of this book and providing incredible feedback. Ari Cohen Goldberg deserves my thanks for providing a great deal of expertise in the realm of language. Thanks to Brian Switek for talking with me about how our knowledge about dinosaurs has changed in the past several decades, making sure I didn't get too much wrong about dinosaurs, and spending longer than anyone else I have spoken with reminiscing about the brontosaurus. In addition, both Brian and Ari were instrumental in shaping my thinking about the generational component of knowledge change, each independently bringing it to my attention. Sarah Gilbert and Rena Lauer were also instrumental in helping with my questions about medieval Europe.

Countless people have also pointed me to articles and ideas, and supported me in numerous other ways during this process. Thank you to everyone.

Of course, if you've gotten this far, you know that facts become out-of-date and errors propagate. Therefore, it's only natural for me to inform the reader that any errors are, of course, my own. You have been so warned.

I owe a great deal of appreciation to Max Brockman, my agent, for having brought me into the world of book writing and providing advice throughout. Courtney Young, my editor, shepherded this book at every stage and has earned my perpetual gratitude. And Niki Papadopoulos, you joined the editing process at the very end but have been a great help in making sure that the book ended up being the best version possible, and actually got it out the door.

My parents have been supportive of this project all along, having read multiple drafts, for which I am incredibly appreciative. But more important, they instilled in me a love of learning. I have strived to live by their daily exhortation to me before heading off to elementary school: "Think, have fun, and be a mensch."

My grandfather, Irwin Arbesman, in addition to allowing me to kick off the book with a great story, is an amazing sounding board for all of my ideas, and has provided wonderful feedback during this entire process. I owe him a great deal of thanks.

And, last, I'd like to thank my wife, Debra. She has been incredibly supportive and proud of me at every stage. She read so many drafts, giving comments on each, and has been willing to listen to me speak about the topics in this book over and over, ad nauseam (at least for her). Debra, you are truly an *eishet chayil*.

NOTES

CHAPTER 1: THE HALF-LIFE OF FACTS

1   **"the diploid chromosome number of 48 in man":** Martin, Aryn. "Can't Any Body Count? Counting as an Epistemic Theme in the History of Human Chromosomes." *Social Studies of Science* 34, no. 6 (December 1, 2004): 923–48; Tjio, Joe Hin and Albert Levan. "The Chromosome Number of Man." *Hereditas* 42, no. 1–2 (1956): 1–6.

1   **But in 1956, Joe Hin Tjio and Albert Levan:** Gartler, Stanley M. "The Chromosome Number in Humans: A Brief History." *Nature Reviews Genetics* 7, no. 8 (August 2006): 655–60.

5   **Certain fields use fact to mean an objective truth:** Philosophers of science will no doubt view my definition of facts and knowledge as distinct from underlying truth as overly simplistic. However, I do not mean that they are separate from some objective scientific truth, simply that they are approaching this truth, as I make clear. In addition, I adhere to the perspective that such an objective scientific truth does exist independent of our minds, and I am quite optimistic that we can move toward it. But, as I discuss further, lots of different types of knowledge change in similar ways, and it is therefore a powerful technique to view them all jointly.

5   **by bundling all of these types of facts together:** A fact and how it changes is ultimately about people: we learn about it from others; we discover it, often by choosing what we wish to explore; and sometimes it is true only because of others.

So let's classify the ways that knowledge changes into four rough categories:

1. What we, as a society, know about the world can be updated.
2. What is true of the world can itself change.

3. As an individual, we can update what we know.
4. As a smaller group of individuals, we can update what we know.

For example, the correction of the number of human chromosomes is an example of the first category. How many billions of people are on the planet is in the second category, as is which computer is the world's most powerful. The third category is simply how we assimilate the first and second categories, sometimes with delays of years or decades, such as in the case of the existence (or not) of the brontosaurus. The fourth category is about how groups of people receive information as it spreads over time, such as when we learn something new, perhaps through the grapevine.

Of course, these are not particularly clear or distinct. Often they are intertwined. For example, the brontosaurus is a little bit of numbers one and three, and even some of four (paleontologists as a group changed faster than the general populace). Which areas of the world were infected with the Black Death involved numbers two and four. And my surprise at discovering that we will hit seven billion people on the planet by the end of 2011 is a combination of numbers two and three.

6   **which I call *mesofacts*:** Mesofacts make up what has been called the *invisible present*:

> All of us can sense change—the reddening sky with dawn's new light, the rising strength of lake waves during a thunderstorm, and the changing seasons of plant flowering as temperature and rain affect our landscapes. Some of us see longer-term events and remember that there was less snow last winter or the fishing was better a couple of years ago. But it is the unusual person who senses with any precision changes occurring over decades. At this timescale, we are inclined to think the world is static, and we typically underestimate the degree of change that does occur. Because we are unable directly to sense slow changes, and because we are even more limited in our abilities to interpret their cause-and-effect relations, processes acting over decades are hidden and reside in what I call "the invisible present." (Magnuson, John J. "Long-term Ecological Research and the Invisible Present." *Bioscience* 40 [1990]: 495–501.)

CHAPTER 2: THE PACE OF DISCOVERY

9   **When Derek J. de Solla Price arrived:** Garfield, Eugene. "A Tribute to Derek John de Solla Price: A Bold, Iconoclastic Historian of

Science." In *Essays of an Information Scientist*, ISI Press. Vol. 7, p. 213.

12   Price published his findings: Price, Derek J. de Solla. "Quantitative Measures of the Development of Science." *Archives Internationales d'Histoire des Sciences* 4, no. 14 (1951): 85–93.

13   *Little Science, Big Science:* Price, Derek J. de Solla. *Little Science, Big Science—and Beyond.* New York: Columbia University Press, 1986.

13   Harvey Lehman published a curious little paper: Lehman, Harvey C. "The Exponential Increase of Man's Cultural Output." *Social Forces* 25, no. 3 (March 1, 1947): 281–90.

14   along with a few more recent areas examined: Enquist, M., et al. "Why Does Human Culture Increase Exponentially?" *Theoretical Population Biology* 74 (2008): 46–55.

16   A group of researchers at Harvard Medical School: Lee, Kyungjoon, John S. Brownstein, Richard G. Mills, and Isaac S. Kohane. "Does Collocation Inform the Impact of Collaboration?" *PLoS ONE* 5, no. 12 (December 15, 2010): e14279.

16   a group of researchers at Northwestern University: Wuchty, Stefan, et al. "The Increasing Dominance of Teams in Production of Knowledge." *Science* 316, no. 5827 (May 18, 2007): 1036–39.

17   It was created by Jorge Hirsch: Hirsch, Jorge E. "An Index to Quantify an Individual's Scientific Research Output." *Proceedings of the National Academy of Sciences of the United States of America* 102 , no. 46 (November 15, 2005): 16569–72.

17   The National Science Foundation has examined how much money: Lehrer, Jonah. "Fleeting Youth, Fading Creativity." *Wall Street Journal*, February 19, 2010.

18   decided to study the scientific output of Nobel laureates: Zuckerman, Harriet. "Nobel Laureates in Science: Patterns of Productivity, Collaboration, and Authorship." *American Sociological Review* 32, no. 3 (1967): 391–403.

20   the distorted words we often have to read correctly: This is known as the reCAPTCHA project and can be found here: www .google.com/recaptcha.

21   *eurekometrics:* Arbesman, Samuel, and Nicholas A. Christakis. "Eurekometrics: Analyzing the Nature of Discovery." *PLoS Computational Biology* 7, no. 6 (June 2011): e1002072.

23   "In the 1940s there are six such moments": Cowen, Tyler. "The Great Stagnation in Medicine." *Marginal Revolution*, 2011. www.marginalrevolution.com/marginalrevolution/2011/02/the -great-stagnation-in-medicine.html.

23   a Swedish medical student named Ivar Sandström: Carney, J. Aidan. "The Glandulae Parathyroideae of Ivar Sandström: Contributions

from Two Continents." *American Journal of Surgical Pathology* 20, no. 9 (1996): 1123–44.

24   if you uttered the statement: Price. *Little Science, Big Science.*

CHAPTER 3: THE ASYMPTOTE OF TRUTH

26   Marjorie Courtenay-Latimer, a young woman in South Africa: Goodall, Jane, Gail Hudson, and Thane Maynard. *Hope for Animals and Their World: How Endangered Species Are Being Rescued from the Brink.* New York: Grand Central Publishing, 2009.

27   an example of what are known as Lazarus taxa: Intriguingly, since the discovery of living coelacanths, fossil coelacanths have been found from the past sixty-five million years. Personal communication with Brian Switek.

28   two biologists at the University of Queensland in Australia: Fisher, Diana O., and Simon P. Blomberg. "Correlates of Rediscovery and the Detectability of Extinction in Mammals." *Proceedings of the Royal Society B: Biological Sciences* (September 29, 2010).

28   a team of scientists at a hospital in Paris: Poynard, Thierry, et al. "Truth Survival in Clinical Research: An Evidence-Based Requiem?" *Annals of Internal Medicine* 136, no. 12 (2002): 888–95.

28   a clear decay in the number of papers that were still valid: This decay is a relatively smooth linear decline, so we might need many more years of data to fit it nicely to an exponential decay.

31   towering well above other publications: Sometimes they become so important that they actually stop being cited. Newton's work is foundational for physics to such a large degree that it has become unnecessary to cite his books.

32   a study of all the papers in the Physical Review journals: Redner, Sidney. "Citation Statistics from More Than a Century of Physical Review" (2004). http://arxiv.org/abs/physics/0407137.

32   Other researchers have even broken this down by subfield: Midorikawa, N. "Citation Analysis of Physics Journals: Comparison of Subfields of Physics." *Scientometrics* 5, no. 6 (November 26, 1983): 361–74.

32   In medicine: Tonta, Yaşar, and Yurdagül Ünal. "Scatter of Journals and Literature Obsolescence Reflected in Document Delivery Requests." *Journal of the American Society for Information Science and Technology* 56, no. 1 (2005): 84–94; doi:10.1002/asi.20114.

32   Price himself examined journals from different fields: Price, Derek J. de Solla. "Citation Measures of Hard Science, Soft Science, Technology, and Nonscience." In *Communication Among*

Scientists and Engineers, eds. C. E. Nelson and D. K. Pollock. Lexington, MA: Heath, 1970. pp. 3–22.

32   **Rong Tang looked at scholarly books in different fields:** Tang, Rong. "Citation Characteristics and Intellectual Acceptance of Scholarly Monographs." *College Research Libraries* 69, no. 4 (2008): 356–69.

33   **three scientists working at the Thermophysical Research Properties Center at Purdue University:** Ho, C. Y., R. W. Powell, and P. E. Liley. "Thermal Conductivity of the Elements." *Journal of Physical and Chemical Reference Data* 1, no. 2 (April 1972): 279–421.

35   **Isaac Asimov, in a wonderful essay:** Asimov, Isaac. "The Relativity of Wrong." *The Skeptical Inquirer* 14, no. 1 (1989): 35–44.

36   **Sean Carroll . . . wrote a wonderful series on his blog:** Carroll, Sean. "The Laws Underlying the Physics of Everyday Life Are Completely Understood." *Cosmic Variance*, 2010; http://blogs.discovermagazine.com/cosmicvariance/2010/09/23/the-laws-underlying-the-physics-of-everyday-life-are-completely-understood.

37   **Carroll even lays down, in a single equation:** This is known as the Dirac equation. Carroll, Sean. "Physics and the Immortality of the Soul." *Cosmic Variance*, 2010; http://blogs.discovermagazine.com/cosmicvariance/2011/05/23/physics-and-the-immortality-of-the-soul/.

37   **A quote from *Science Daily*:** Census of Marine Life. "Giant Undersea Microbial Mat Among Discoveries Revealed by Marine Life Census." *Science Daily*, April 18, 2010.

38   **Kevin Kelly refers to this sort of distribution:** Kelly, Kevin. "The Long Tail of Life." *The Technium*, 2010; http://www.kk.org/thetechnium/archives/2010/04/the_long_tail_o.php.

CHAPTER 4: MOORE'S LAW OF EVERYTHING

41   **The @ symbol has been on keyboards:** Rawsthorn, Alice. "Why @ Is Held in Such High Design Esteem." *International Herald Tribune*, March 22, 2010.

42   **Moore wrote a short paper in the journal *Electronics*:** Moore, Gordon E. "Cramming More Components Onto Integrated Circuits" (Reprinted from *Electronics*, pg 114–17, April 19, 1965). *Proceedings of the IEEE* 86, no. 1 (1998): 82–85.

43   **number of pixels that digital cameras can process:** Myhrvold, Nathan. "Moore's Law Corollary: Pixel Power." *New York Times*, June 7, 2006.

43   **Magee, along with a postdoctoral fellow Heebyung Koh:** Koh, Heebyung, and Christopher L. Magee. "A Functional Approach for Studying Technological Progress: Application to Information

Technology." *Technological Forecasting and Social Change* 73, no. 9 (2006): 1061–83; Koh, Heebyung, and Christopher L. Magee. "A Functional Approach for Studying Technological Progress: Extension to Energy Technology." *Technological Forecasting and Social Change* 75, no. 6 (2008): 735–58.

45   **They begin to run out of space:** This growth pattern assumes a continuous food supply.

45   **Clayton Christensen, a professor at Harvard Business School:** Christensen, Clayton M. "Exploring the Limits of the Technology S-Curve. Part I: Component Technologies." *Production and Operations Management* 1, no. 4 (1992): 334–57.

46   **they found mathematical regularities:** More recent research has debated whether these are truly exponential or other fast-growing functions, such as power laws or double exponentials. The upshot is the same: There are regularities. See McNerney, James, et al. "Role of Design Complexity in Technology Improvement." *Proceedings of the National Academy of Sciences* 108, no. 22 (May 31, 2011): 9008–13; Nagy, Béla, et al. "Superexponential Long-term Trends in Information Technology." *Technological Forecasting and Social Change* 78, no. 8 (October 2011): 1356–64. For more curves, see the performance curve database. http://pcdb.santafe.edu.

46   **has even been found in robots:** Powell, Corey S. "The Rise of the Machines Is Not Going as We Expected." *Discover* (May 2010).

46   **Kevin Kelly, in his book *What Technology Wants:*** Kelly, Kevin. *What Technology Wants.* New York: Viking, 2010.

48   **If you plot prefix sizes against the years:** Personal research. Underlying data from Bureau International des Poids et Mesures. *The International System of Units,* 2006.

48   **cost of genome sequencing is dropping rapidly:** MacArthur, Daniel. "The Plummeting Cost of Genome Sequencing." *Wired Science,* 2011. http://www.wired.com/wiredscience/2011/02/illustrating-the-plummeting-cost-of-genome-sequencing.

48   **These technological doublings in the realm of science:** Mathematics also has such doublings. For example, the number of digits of the largest known prime number has increased exponentially, due to growth in computing power that allows for the discovery of larger and larger primes. Further information available here: http://primes.utm.edu/notes/by_year.html

48   **Moore's Law of proteomics:** Cox, Jürgen, and Matthias Mann. "Is Proteomics the New Genomics?" *Cell.* Cell Press, August 10, 2007. http://linkinghub.elsevier.com/retrieve/pii/S0092867407009701.

48   **the number of neurons that can be:** Stevenson, Ian H., and Konrad P. Kording. "How Advances in Neural Recording Affect Data Analysis." *Nature Neuroscience* 14, no. 2 (February 2011): 139–42.

49 "Science and technology are closely related": Cole, Jonathan R. *The Great American University: Its Rise to Preeminence, Its Indispensable National Role, Why It Must Be Protected.* New York: Public Affairs, 2009.

49 Henry Petroski, a professor of engineering and history: Petroski, Henry. "Engineering Is Not Science." *IEEE Spectrum* 9 December 2010.

50 In 1928, the engineer Trygve Dewey Yensen: Yensen, T. D. "What Is the Magnetic Permeability of Iron?" *Journal of the Franklin Institute* 206, no. 4 (1928): 503–10.

51 game after game has become: Cirasella, J. and D. Kopec. "The History of Computer Games." Exhibit at Dartmouth Artificial Intelligence Conference: The Next Fifty Years (AI@50). Conferences, Seminars and Symposiums: Conference Presentation. Dartmouth College, Hanover, NH, July 13–15, 2006.

52 polio is now generally regarded: Unfortunately, polio's eradication in the developed world is one of our more slippery mesofacts. Polio still exists in developing countries, and due to our globalized world, it still has the potential of spreading to developed countries where it has been eliminated.

53 while aluminum used to be the most valuable: Kotler, Steven, and Peter H. Diamandis. *Abundance: The Future Is Better Than You Think.* New York: Free Press, 2012.

53 have added about 0.4 years: Wolfram|Alpha. "Life Expectancy, United States"; http://www.wolframalpha.com/input/?i=life+expectancy+United+States, 2011.

53 *actuarial escape velocity:* Grey, Aubrey D. N. J. de. "Escape Velocity: Why the Prospect of Extreme Human Life Extension Matters Now." *PLoS Biol* 2, no. 6 (June 15, 2004): e187. Further reading: Finch, Caleb E., and Eileen M. Crimmins. "Inflammatory Exposure and Historical Changes in Human Life-Spans." *Science* 305, no. 5691 (September 17, 2004): 1736–39.

55 The physicist Tom Murphy has shown: Murphy, Tom. "Galactic-Scale Energy." *Do the Math*, 2011. http://physics.ucsd.edu/do-the-math/2011/07/galactic-scale-energy.

55 self-fulfilling propositions: Kelly. *What Technology Wants.*

56 The Hawthorne effect was defined: McCarney, Rob, et al. "The Hawthorne Effect: A Randomised, Controlled Trial." *BMC Medical Research Methodology* 7, no. 1 (2007): 30.

57 The tiny population of Tasmania: Henrich, Joseph. "Demography and Cultural Evolution: How Adaptive Cultural Processes Can Produce Maladaptive Losses: The Tasmanian Case." *American Antiquity* 69, no. 2 (April 1, 2004): 197–214.

58 "The more populous periods": Caplan, Bryan. http://econlog.econlib.org/archives/2011/05/replies_to_crit.html. 2011.

58   A classic paper by economist Michael Kremer: Kremer, Michael. "Population Growth and Technological Change: One Million B.C. to 1990." *The Quarterly Journal of Economics* 108, no. 3 (August 1, 1993): 681–716.

59   More recent research: Bettencourt, L. M. A., J. Lobo, D. Helbing, C. Kuhnert, and G. B. West. "Growth, Innovation, Scaling, and the Pace of Life in Cities." *Proceedings of the National Academy of Sciences of the United States of America* 104, no. 17 (2007): 7301–06.

59   Using these two assumptions: Kremer. "Population Growth."

60   *first-order model:* The very simplest model is a zeroth-order model, but here we have at least the relationship between population and technological progress.

61   higher population densities in certain regions: Ashraf, Quamrul, and Oded Galor. "Dynamics and Stagnation in the Malthusian Epoch." *National Bureau of Economic Research Working Paper Series* no. 17037 (2011).

61   the concerns of the English people: Merton, Robert K. "Science, Technology and Society in Seventeenth-Century England." *Osiris* 4 (January 1, 1938): 360–632.

62   decided in 1989 to make a special sort of map: Cliff, Andrew, and Peter Haggett. "Time, Travel and Infection." *British Medical Bulletin* 69 (January 2004): 87–99.

63   these other modes of transportation: Ibid.; Grübler, Arnulf. *Technology and Global Change.* Cambridge, UK: Cambridge University Press. 2003.

64   examined the city of Berlin: Marchetti, Cesare. "Anthropological Invariants in Travel Behavior." *Technological Forecasting and Social Change* 47, no. 1 (September 1994): 75–88.

64   The Black Death spread: Noble, J. V. "Geographic and Temporal Development of Plagues." *Nature* 250, no. 5469 (August 30, 1974): 726–29.

CHAPTER 5: THE SPREAD OF FACTS

66   a group of telephone interviewers: Schwartz, David A. "How Fast Does News Travel?" *The Public Opinion Quarterly* 37, no. 4 (December 1, 1973): 625–27.

67   Consider the case of Mary Tai: Tai, Mary M. "A Mathematical Model for the Determination of Total Area Under Glucose Tolerance and Other Metabolic Curves." *Diabetes Care* 17, no. 2 (February 1, 1994): 152–54.

71   how certain cities were affected: Dittmar, Jeremiah E. "Information Technology and Economic Change: The Impact of the Printing Press." *The Quarterly Journal of Economics* 126, no. 3 (August 1, 2011): 1133–72.

71 **Gutenberg combined and extended a whole host of technologies:** I recommend going to the Gutenberg Museum in Mainz, which goes into all of this in astonishing depth.

74–75 **we have measured the average number of close social connections:** Christakis, Nicholas A., and James H. Fowler. *Connected: The Surprising Power of Our Social Networks and How They Shape Our Lives*. New York: Little Brown, 2009.

75 **We understand how social groups are distributed:** Onnela, Jukka-Pekka, et al. "Geographic Constraints on Social Network Groups." *PLoS ONE* 6, no. 4 (April 5, 2011): e16939.

75 **work done by him and his longtime collaborator James Fowler:** Christakis and Fowler. *Connected*. The methodology in this research has been critiqued more recently. For details, see the following papers: VanderWeele, T. J. "Sensitivity Analysis for Contagion Effects in Social Networks." *Sociological Methods and Research* 40 (2011): 240–55; Christakis, Nicholas A., and James H. Fowler. "Social Contagion Theory: Examining Dynamic Social Networks." *Statistics in Medicine*, forthcoming.

81 **"[c]onclusions based on such work":** Gould, Stephen Jay. *Bully for Brontosaurus: Reflections in Natural History*. New York: W. W. Norton, 1992.

83–84 **the error was so widespread:** Hamblin, T. J. "Fake!" *British Medical Journal* 283 (1981): 19–26. The persistence of incorrect facts in the literature over time was explored further in Tatsioni, Athina, Nikolaos G. Bonitsis, and John P. A. Ioannidis. "Persistence of Contradicted Claims in the Literature." *JAMA: The Journal of the American Medical Association* 298, no. 21 (December 5, 2007): 2517–26.

84 **While working on his book:** Mauboussin, Michael J. *Think Twice: Harnessing the Power of Counterintuition*. Harvard Business School Press, 2009; *See For Yourself: The Importance of Checking Claims*, Legg Mason Global Asset Management, 2009.

84 **author Randall Munroe wishes for a world:** Munroe, Randall. "Misconceptions." *xkcd*. https://www.xkcd.com/843/.

85 **I have found instances of it:** For these and more, search for the phrase *contrary to popular belief* in Google Books.

85 **"Dynamics of an asteroid":** Bothamley, Jennifer, ed. *Dictionary of Theories*. Farmington Hills, MI: Gale Research International Ltd., 1993.

85 **the article referenced in *New Scientist*:** Bowers, John F. "James Moriarty: A Forgotten Mathematician." *New Scientist* (December 23–30, 1989).

86 **But the citations to Moriarty's work:** Kennaway, K. D. "String Theory and the Vacuum Structure of Confining Gauge Theories." PhD dissertation. University of Southern California, 2004.

86   These errors "will live on and on": Mauboussin, *See For Yourself*.

86   I was taken to task soon after by James Fallows: Fallows, James. "Boiled Frog Does a Surreal Meta-Backflip." *The Atlantic*, March 2, 2010. http://www.theatlantic.com/technology/archive/ 2010/03/boiled-frog-does-a-surreal-meta-backflip/36934/.

90   research that quantitatively studied the differences: Barbrook, Adrian C. et al. "The Phylogeny of *The Canterbury Tales*." *Nature* 394, no. 6696 (August 27, 1998): 839.

91   actually measured: Among others, here are a couple of their papers: Simkin, M. V., and V. P. Roychowdhury. "Stochastic Modeling of Citation Slips." *Scientometrics* 62 (2005): 367–84; "Read Before You Cite!" arXiv:cond-mat/0212043 (December 2002).

92   From their paper: Liben-Nowell, David, and Jon Kleinberg. "Tracing Information Flow on a Global Scale Using Internet Chain-letter Data." *Proceedings of the National Academy of Sciences* 105, no. 12 (March 25, 2008): 4633–38.

93   it is often the case that credible information or news spreads faster: Castillo, Carlos, Marcelo Mendoza, and Barbara Poblete. "Information Credibility on Twitter." In *Proceedings of the 20th International Conference on World Wide Web*. New York: Association for Computing Machinery, 2011. 675–84

CHAPTER 6: HIDDEN KNOWLEDGE

96   increased use of antibacterial soaps: Arbesman, Harvey. "Is Cutaneous Malignant Melanoma Associated with the Use of Antibacterial Soaps?" *Medical Hypotheses* 53, no. 1 (July 1999): 73–75.

96   dairy consumption is related to acne: Arbesman, Harvey. "Dairy and Acne—the Iodine Connection." *Journal of the American Academy of Dermatology* 53(6): 1102. December 2005.

97   when Roche brought a problem: "The Benefits of Open Innovation." InnoCentive. http://www.innocentive.com/seekers/benefits -open-innovation.

97   When NASA used InnoCentive: Legatum Center for Development & Entrepreneurship. "Legatum Lecture Series Presents: Alpheus Bingham of InnoCentive, Inc." http://legatum.mit.edu/ binghamlecture.

99   He demonstrated this with a novel finding: Swanson, Don R. "Undiscovered Public Knowledge." *The Library Quarterly* 56, no. 2 (April 1, 1986): 103–18.

99   Building on this, Swanson continued: Swanson, Don R. "Medical Literature as a Potential Source of New Knowledge." *Bulletin of the Medical Library Association* 78, no. 1 (January 1990):

29–37; Swanson, Don R. "Migraine and Magnesium: Eleven Neglected Connections." *Perspectives in Biology and Medicine* 31, no. 4 (1988): 526–57.

100 **revisited undiscovered public knowledge:** Swanson, Don R., and Neil Smalheiser. "Undiscovered Public Knowledge: A Ten-Year Update." In *KDD-96 Proceedings*. Edited by Evangelos Simoudis, Jia Han, and Usama Fayyad, 295–98. AAAI Press, 1996.

102 **has a greater chance of solving a problem than do the experts:** About one-third of all InnoCentive challenges yield solutions. Lakhani, K. R., et al. "The Value of Openness in Scientific Problem Solving." Harvard Business School Working Paper No. 07—050. (2007); http://www.hbs.edu/research/pdf/07—050.pdf.

103 **the same scientists who explored the errors:** Simkin, M. V., and V. P. Roychowdhury. "Re-inventing Willis." *Physics Reports* 502, no. 1 (May 2011): 1–35.

104 **Certain concepts in computer science:** Trakhtenbrot, B. A. "A Survey of Russian Approaches to Perebor (Brute-Force Searches) Algorithms." *IEEE Annals of the History of Computing* 6 (October 1, 1984): 384–400.

107 **how often scientists were aware of previous research:** Robinson, Karen A., and Steven N. Goodman. "A Systematic Examination of the Citation of Prior Research in Reports of Randomized, Controlled Trials." *Annals of Internal Medicine* 154, no. 1 (January 4, 2011): 50–55.

108 **a team of scientists from the hospitals and schools:** Lau, Joseph, et al. "Cumulative Meta-Analysis of Therapeutic Trials for Myocardial Infarction." *New England Journal of Medicine* 327, no. 4 (July 23, 1992): 248–54.

110 **the creation of a massive database:** Frijters, Raoul, et al. "Literature Mining for the Discovery of Hidden Connections between Drugs, Genes and Diseases." *PLoS Computational Biology* 6, no. 9 (September 23, 2010): e1000943.

110 **CoPub Discovery involved:** Frijters, Raoul, et al. "CoPub: A Literature-Based Keyword Enrichment Tool for Microarray Data Analysis." *Nucleic Acids Research* 36, no. supplement 2 (July 1, 2008): W406–W410.

111 **CoPub Discovery predicted:** Another example of a tool like this: Kuhn, Michael, et al. "Large-Scale Prediction of Drug-Target Relationships." *Federation of European Biochemical Societies Letters* 582, no. 8 (April 9, 2008): 1283–90.

112 **software designed to find undiscovered patterns:** See TRIZ, a method of invention and discovery. For example, here: www.aitriz.org.

112 **computerized systems devoted to drug repurposing:** Sanseau, Philippe, and Jacob Koehler. "Editorial: Computational Methods for Drug Repurposing." *Briefings in Bioinformatics* 12, no. 4 (July 1, 2011): 301–2.

112 **can generate new and interesting:** Darden, Lindley. "Recent Work in Computational Scientific Discovery." In *Proceedings of the Nineteenth Annual Conference of the Cognitive Science Society* (1997) 161–66.

113 **names a novel, computationally created:** See TheoryMine: http://theorymine.co.uk.

116 **A Cornell professor of earth and atmospheric sciences:** Cisne, John L. "How Science Survived: Medieval Manuscripts' 'Demography' and Classic Texts' Extinction." *Science* 307, no. 5713 (February 25, 2005): 1305–7.

119 **"This can create almost lyrical connections":** Johnson, Steven Berlin. "Tool for Thought." *New York Times*, January 30, 2005.

CHAPTER 7: FACT PHASE TRANSITIONS

121 **Thomas Wright, a British astronomer:** Sagan, Carl. *The Varieties of Scientific Experience.* New York: Penguin, 2006.

127 **In 1953:** Kelly, Kevin. *What Technology Wants.* New York: Viking, 2010. p. 157.

128 **Figuring out the right underlying change to measure:** In physics, this is related to finding what is known as the *order parameter*, the quantity that is zero in one phase and nonzero in another phase. Determining this requires a certain amount of creative effort for each system.

129 **a stunningly discontinuous jump in our knowledge:** Arbesman, Samuel, and Gregory Laughlin. "A Scientometric Prediction of the Discovery of the First Potentially Habitable Planet with a Mass Similar to Earth." *PLoS ONE* 5, no. 10 (October 2010): e13061.

130 **a simple metric of habitability:** Others have since developed other habitability metrics. See for example, the Earth Similarity Index: http://phl.upr.edu/projects/earth-similarity-index-esi

131 **Kepler 22b:** Kepler 22b is likely quite a bit more massive than Earth. However, there are certain planetary candidates discovered by the Kepler mission that are even more Earth-like but have not been confirmed as of mid-2012.

132 **whole new pieces of math were involved:** Singh, Simon. *Fermat's Enigma.* New York: Walker & Company, 1997.

134 **There are some problems:** This was shown by Kurt Gödel in his Incompleteness Theorem. For further reading, see for example, *Gödel, Escher, Bach: An Eternal Golden Braid* by Douglas R. Hofstadter. Basic Books. 1979.

135   **it turns out we will have to wait until 2024:** Arbesman, Samuel, and Rachel Courtland. "2011 preview: Million-Dollar Mathematics Problem." *New Scientist*, December 2010. More recently, Ryohei Hisano and Didier Sornette have conducted a more sophisticated statistical analysis; they estimate that there's a 41 percent chance by 2024, similar to our prediction of 50 percent. Hisano, Ryohei, and Didier Sornette. "On the Distribution of Time-to-Proof of Mathematical Conjectures." (2012); http://arxiv .org/abs/1202.3936.

135   **Luís Bettencourt and his colleagues:** Bettencourt, Luis M. A., et al. "Growth, Innovation, Scaling, and the Pace of Life in Cities." *Proceedings of the National Academy of Sciences* 104, no. 17 (2007): 7301–6.

136   **the way it does for living things:** West, G. B, J. H. Brown, and B. J. Enquist. "A General Model for the Origin of Allometric Scaling Laws in Biology." *Science* 276 (5309): 126.

137   **there would be little difference:** Personal communication with Sarah Gilbert and Rena Lauer, both medieval historians.

137   **even fashion during the Middle Ages:** Loschek, Ingrid. *When Clothes Become Fashion: Design and Innovation Systems.* London: Berg Publishers, 2009.

CHAPTER 8: MOUNT EVEREST AND THE DISCOVERY OF ERROR

141   **we now know for certain:** See Everest@National Geographic: http://www.nationalgeographic.com/features/99/everest/roof_ content.html (accessed December 20, 2011).

142   **the world record for the tallest tree:** Preston, Richard. "Tall for Its Age." *The New Yorker* (October 9, 2006): 32–36.

143   **"Revolutions in science have often been preceded":** Cukier, Kenneth. "A Special Report on Managing Information: Data, Data Everywhere." *The Economist* (February 25, 2010).

144   **Wilkins went on to define a regular system of lengths:** Wilkins, John. *An Essay Towards a Real Character, and a Philosophical Language.* 1668. Available online: http://www.metricationmat ters.com/docs/WilkinsTranslationShort.pdf

146   **the speed of light and the length of the meter:** For more on measurements, see the Web site of the National Institute of Standards and Technology: http://www.nist.gov/.

146   **The world of measurement involves much more:** Cardarelli, François. *Encyclopaedia of Scientific Units, Weights, and Measures: Their SI Equivalences and Origins.* New York: Springer Publishing, 2004.

148   **have been bandying about alternative definitions:** Crease, Robert P. "Measurement and Its Discontents." *New York Times.* October 23, 2011.

148   **published a tongue-in-cheek paper:** Dessler, A. J., and C. T. Russell. "From the Ridiculous to the Sublime: The Pending Disappearance of Pluto." *Eos, Transactions, American Geophysical Union* 61, no. 44 (1980): 690.

153   **there is about a one in six chance:** The Fisher's exact test was used here. An online calculator is available here: http://www .graphpad.com/quickcalcs/contingency1.cfm

153   **illustrated some of the failings of this threshold:** Munroe, Randall. "Significant." *xkcd.* http://xkcd.com/882/.

154   **"Statistics is the science":** Penman, Bridget, et al. "Genome-wide Association Studies in Plasmodium Species." *BMC Biology* 8, no. 1 (2010): 90; Statisticians have developed techniques to account for this problem, such as the use of something known as the Bonferroni correction. This simply states that if you are testing lots and lots of variables to see if they are related to something in a significant way, what you deem a significant p-value must be much more strict, and much smaller.

155   **Planet X was a slippery thing:** Quinlan, Gerald D. "Planet X: A Myth Exposed." *Nature* 363, no. 6424 (May 6, 1993): 18–19. Grosser, Morton. "The Search for a Planet beyond Neptune." *Isis* 55, no. 2 (June 1, 1964): 163–83. The History of Science Society.

157   **He has found that for highly cited clinical trials:** Ioannidis, John P. A. "Contradicted and Initially Stronger Effects in Highly Cited Clinical Research." *JAMA: The Journal of the American Medical Association* 294, no. 2 (2005): 218–28.

158   **Ioannidis conducted the same test for various biomarkers:** Ioannidis, John P. A., and Orestis A. Panagiotou. "Comparison of Effect Sizes Associated With Biomarkers Reported in Highly Cited Individual Articles and in Subsequent Meta-analyses." *JAMA: The Journal of the American Medical Association* 305, no. 21 (June 1, 2011): 2200–10.

158   **what is perhaps Ioannidis's most well-known paper:** Ioannidis, John P. A. "Why Most Published Research Findings Are False." *PLoS Med* 2, no. 8 (2005): e124.

161   **Regarding a kerfuffle:** Zimmer, Carl. "It's Science, but Not Necessarily Right." *New York Times*, June 26, 2011.

162   **"If it confirmed the first researcher's findings":** Quoted in Cole, Stephen. *Making Science: Between Nature and Society.* Cambridge, MA: Harvard University Press, 1992.

162   **researchers calculated that a small amount of replication:** Moonesinghe, Ramal, Muin J. Khoury, A. Cecile, and J. W. Janssens. "Most Published Research Findings Are False—But a Little

Replication Goes a Long Way." *PLoS Medicine* 4, no. 2 (February 27, 2007): e28.

163   **As Lord Florey, a president of the Royal Society, stated:** Cole, Jonathan, and Stephen Cole. *Social Stratification in Science.* Chicago: University of Chicago, 1973. p. 217.

163   **Science is not always cumulative:** Cole, Stephen. *Making Science: Between Nature and Society.* Cambridge, MA: Harvard University Press, 1992.

164   **"The scientific literature is strewn":** Ziman, John M. *Public Knowledge: An Essay Concerning the Social Dimension of Science.* Cambridge, UK: Cambridge University Press, 1968.

165   **A stark example is that of war:** Mueller, John. "War Has Almost Ceased to Exist: An Assessment." *Political Science Quarterly,* 124, no. 2 (2009).

166   **When born in 1822, Francis Galton:** Galton, Francis. *Inquiries into Human Faculty and Its Development,* 1883. Available online: http://galton.org/books/human-faculty/text/human -faculty.pdf

166   **He wrote a paper:** Galton, Francis. "On Head Growth in Students at the University of Cambridge." *The Journal of the Anthropological Institute of Great Britain and Ireland* 18 (January 1, 1889): 155–56.

166   **how people visualize numbers in their mind:** Galton, Francis. "Visualized Numerals." *Nature* (March 25, 1880): 494–5.

166   **how many pretty women he encountered:** Gorraiz, Juan, Christian Gumpenberger, and Martin Wieland. "Galton 2011 Revisited: A Bibliometric Journey in the Footprints of a Universal Genius." *Scientometrics* 88, no. 2 (2011): 627–52.

166   **Galton was the man who ushered in the Statistical Enlightenment:** Stigler, S. M. "Darwin, Galton and the Statistical Enlightenment." *Journal of the Royal Statistical Society: Series A (Statistics in Society)* 173 (2010): 469–82.

166   **wrote the following of Galton:** Price, Derek J. de Solla. *Little Science, Big Science—and Beyond.* New York: Columbia University Press, 1986.

167   **the elderly are capable of crossing the street:** Hoxie, R. E., and L. Z. Rubenstein. "Are Older Pedestrians Allowed Enough Time to Cross Intersections Safely?" *Journal of the American Geriatrics Society* 42, no. 3 (March 1994): 241–4.

168   **are the subject of the vast majority of scientific papers:** May, Robert M. "How Many Species Are There on Earth?" *Science* 241, no. 4872 (September 16, 1988): 1441–49; Clark, J. Alan, and Robert M. May. "Taxonomic Bias in Conservation Research." *Science* 297, no. 5579 (July 12, 2002): 191–92.

168   some scientists even call it *taxonomic chauvinism:* Bonnet, Xavier, Richard Shine, and Olivier Lourdais. "Taxonomic Chauvinism." *Trends in Ecology & Evolution.* Amsterdam: Elsevier Science Publishers, (January 1, 2002).

169   one of the main reasons that the brontosaurus: Gould, Stephen Jay. *Bully for Brontosaurus: Reflections in Natural History.* New York: W. W. Norton, 1992.

CHAPTER 9: THE HUMAN SIDE OF FACTS

171   Frogs have a curious type of vision: Lettvin J. Y., et al. "What the Frog's Eye Tells the Frog's Brain," *Proceedings of the Institute of Radio Engineers* 47 (1959): 1940–51, reprinted in Warren S. McCulloch, *Embodiments of Mind.* Cambridge, MA: MIT Press, 1965.

173   the abundance of cod: Helfman, Gene S. *Fish Conservation: A Guide to Understanding and Restoring Global Aquatic Biodiversity and Fishery Resources.* Washington, DC: Island Press, 2007.

173   "anything that was invented after you were born": Kelly, Kevin. *What Technology Wants.* New York: Viking, 2010. p. 235.

174   they recanted their editorial: "A Correction." *New York Times,* July 17, 1969.

174   Why do we believe in wrong, outdated facts?: Schulz, Kathryn. *Being Wrong: Adventures in the Margin of Error.* New York: Ecco, 2010. One of the main reasons, Schulz notes, that it is so easy to be wrong is very simple: Being wrong feels a lot like being right.

175   Bradley Wray was preparing his high school students: Wray, Bradley. "Cognitive Bias Song" ; https://www.youtube.com/watch?v=3RsbmjNLQkc.

176   Ignaz Semmelweis argued that the doctors: Hempel, Carl G. *Philosophy of Natural Science.* Englewood Cliffs, NJ: Prentice-Hall, 1966.

177   akin to Daniel Kahneman's idea of *theory-induced blindness*: Shirky, Clay. *Cognitive Surplus: Creativity and Generosity in a Connected Age.* New York: The Penguin Press, 2010. p. 99.

178   A series of seminal experiments were done in this field: Chabris, Christopher, and Daniel Simons. *The Invisible Gorilla: And Other Ways Our Intuitions Deceive Us.* New York: Crown Archetype, 2010.

179   Obama decided to sign the guestbook: Amira, Dan. "President Obama Has No Idea What Year It Is." *New York: Daily Intel,* 2011; http://nymag.com/daily/intel/2011/05/president_obama_has_no_idea_wh.html.

181   **according to WebMD:** WebMD. "Blood Nose (Nosebleed) Causes and Treatments"; http://firstaid.webmd.com/nosebleeds -causes-and-treatments; accessed February 4, 2012.

183   **or distinguish Pluto in some other way:** The son of a friend of mine explained to me that Pluto was destroyed, the same way that Superman's home planet of Krypton was destroyed.

184   **reading an essay by Michael Chabon:** Chabon, Michael. "To The Legoland Station." In *Manhood for Amateurs*. New York: HarperCollins, 2009, pp. 51–58.

185   **enriched by spirited discussion:** Johnson, Steven. *Where Good Ideas Come From: The Natural History of Innovation*. New York: Riverhead, 2010.

186   **Pritchett recently proposed an intriguing idea:** Howley, Kerry. "Welcome Guest Workers." *The Atlantic* (July/August 2009).

186   **Kuhn argued that switching from one paradigm to another:** Kuhn, Thomas S. *The Structure of Scientific Revolutions*. Chicago: University of Chicago Press, 1996. p. 151.

188   **Planck's Principle doesn't hold:** Hull, David L. *Science as a Process*. Chicago: University of Chicago Press. 1988; Wray, K. Brad. *Kuhn's Evolutionary Social Epistemology*. Cambridge: Cambridge University Press. 2011.

188   **Watts has demonstrated:** For an overview of his work and this topic, see Watts, Duncan. *Everything Is Obvious *Once We Know the Answer: How Common Sense Fails Us*. New York: Crown Business, 2011.

189   **there is evidence that the frequencies:** Martin, Andrew Thomas. "The Evolving Lexicon." Dissertation. University of California Los Angeles, 2007.

189   **the rate of a verb's regularization:** Lieberman, Erez, et al. "Quantifying the Evolutionary Dynamics of Language." *Nature* 449, no. 7163 (2007): 713–16.

189   **to continue to correct everyone:** For many more examples, see Ben Yagoda's article in *Slate*, "The 'Nonplussed' Problem"; http://www.slate.com/articles/life/the_good_word/2011/04/the_ nonplussed_problem.2.html.

190   **a longer voice onset time:** Kuniko, Nielsen. "Specificity and Abstractness of VOT Imitation." *Journal of Phonetics* 39, no. 2 (April 2011): 132–42.

190   **A team of linguists studied Oprah:** Hay, Jennifer, Stefanie Jannedy, and Norma Mendoza-Denton. "Oprah and /ay/: Lexical Frequency, Referee Design and Style." In *Proceedings of the 14th International Congress of Phonetic Sciences*, eds. John J. Ohala et al. (1999): 1389–92.

191   how this linguistic change happens around us: McWhorter, John. "Swearing In: Are Curse Words Becoming More Common?" *The New Republic*, March 23, 2011.

193   a regional accent based on age: See work by Suzanne Evans Wagner; for example: Wagner, Suzanne Evans. "Language Change and Stabilization in the Transition from Adolescence to Adulthood." Dissertation. University of Pennsylvania, 2008.

194   a certain situational aspect to the shift: Yaeger-Dror, Malcah. "Phonetic Evidence for the Evolution of Lexical Classes: The Case of a Montreal French Vowel Shift." In *Towards a Social Science of Language*, ed. G. Guy, et al. Philadelphia: Benjamins, 1996. 263–87; Yaeger-Dror, Malcah, "Lexical Classes in Montreal French: The Case of (E:)," *Language and Speech* 35 no. 3 (July/September 1992): 251.

195   there is a Web site called Worldometers: http://www.worldometers.info.

197   the Web site MeasuringWorth.com: http://www.measuringworth.com/ppoweruk.

198   a series called Media Diet: http://www.theatlanticwire.com/posts/media-diet.

198   This is already happening: Sparrow, Betsy, Jenny Liu, and Daniel M. Wegner. "Google Effects on Memory: Cognitive Consequences of Having Information at Our Fingertips." *Science* 353, no. 6043 (2011): 776–78.

198   While this is certainly a common argument: Nicholas Carr discusses this topic, in a qualified manner, in his article in the July/August 2008 issue of *The Atlantic*, "Is Google Making Us Stupid?"

198   a constantly updated online medical reference: http://www.uptodate.com/home/about/index.html.

CHAPTER 10: AT THE EDGE OF WHAT WE KNOW

200   This error-checking methodology: Johnson, Steven Berlin. *Where Good Ideas Come From: The Natural History of Innovation*. New York: Riverhead, 2010.

200   the modern conception of the fact: Poovey, Mary. *A History of the Modern Fact: Problems of Knowledge in the Sciences of Wealth and Society*. Chicago: University of Chicago Press, 1998.

201   detailed a number of facts about the origins of human beings: Barnes, Jonathan. *Early Greek Philosophy*. New York: Penguin, 1987.

201   Science requires an idea to be refutable: This is the idea of falsifiability of Karl Popper: A scientific theory is only truly a theory if

it is testable, and can be refuted, or falsified, by contrary evidence. He discusses this in the book, *The Logic of Scientific Discovery*. Routledge. Reprinted in 1992.

201   **"This is the pivotal insight of the Scientific Revolution":** Schulz, Kathryn. *Being Wrong: Adventures in the Margin of Error*. New York: Ecco, 2010. p. 32.

202   **a team of researchers compiled a list:** Deutsch, Karl W., John Platt, and Dieter Senghaas. "Conditions Favoring Major Advances in Social Science." *Science* 171, no. 3970 (February 5, 1971): 450–59.

203   **Do submerged islands . . . remain nation-states:** "I Am a Rock, I Am an Island: How Submerged Islands Could Keep Their Statehood." *The Economist*, May 26, 2011.

203   **there are many who feel:** "Tech Luminaries Address Singularity." *IEEE Spectrum*, June 2008.

204   **its development has gone hand in hand:** This is known as the *demographic transition*.

204   **his taxonomy had three kingdoms:** Natural History Museum, London. "Carl Linnaeus." http://www.nhm.ac.uk/nature-online/science-of-natural-history/biographies/linnaeus/index.html.

205   **the International List of Causes of Death was first adopted:** World Health Organization. "History of the Development of the ICD." Available online: www.who.int/entity/classifications/icd/en/HistoryOfICD.pdf

205   **we are up to the tenth revision:** The American version even has tens of thousands more classifications than the international version.

205   **Just as being exposed:** Johnson, Steven. *Everything Bad Is Good for You*. New York: Riverhead Books, 2005.

205   **This is about the number of soldiers:** Christakis, Nicholas A., and James H. Fowler. *Connected: The Surprising Power of Our Social Networks and How They Shape Our Lives*. New York, New York, USA: Little Brown, 2009.

206   **and is about 190, as of 2011:** Ugander, Johan et al. "The Anatomy of the Facebook Social Graph"; http://arxiv.org/abs/1111.4503.

206   **we increase the number of people we are close to:** O'Malley, A. James, et al. "Egocentric Social Network Structure, Health, and Pro-Social Behaviors in a National Panel Study of Americans." *PLoS ONE*. 7(5): e36250.

206   **Sherlock Holmes argued this very point:** Doyle, Arthur Conan. *A Study in Scarlet*, 1887. First published by Ward Lock & Co. in *Beeton's Christmas Annual*, London. Available online: http://www.gutenberg.org/ebooks/244.

207 **decided to use history as a guide:** Magee, Christopher L., and Tessaleno C. Devezas. "How Many Singularities Are Near and How Will They Disrupt Human History?" *Technological Forecasting and Social Change* 78, no. 8 (October 2011): 1365–78.

208 **"Seriously, the world is changing so quickly":** Flood, Alison. "Jonathan Franzen Warns Ebooks Are Corroding Values." *The Guardian*. January 30, 2012. http://www.guardian.co.uk/books/2012/jan/30/jonathan-franzen-ebooks-values.

INDEX

HILLSBORO PUBLIC LIBRARIES
Hillsboro, OR
Member of Washington County
COOPERATIVE LIBRARY SERVICES